D0806167

INTERNATIONAL SERIES OF MONOGRAPHS IN PURE AND APPLIED BIOLOGY

Division: **ZOOLOGY**

General Editor: G. A. Kerkut

Volume 49

PHYSIOLOGY OF ECHINODERMS

OTHER TITLES IN THE ZOOLOGY DIVISION

General Editor: G. A. KERKUT

Vol. 1. RAVEN—*An Outline of Developmental Physiology*
Vol. 2. RAVEN—*Morphogenesis: The Analysis of Molluscan Development*
Vol. 3. SAVORY—*Instinctive Living*
Vol. 4. KERKUT—*Implications of Evolution*
Vol. 5. TARTAR—*The Biology of Stentor*
Vol. 6. JENKIN—*Animal Hormones—A Comparative Survey. Part 1. Kinetic and Metabolic Hormones*
Vol. 7. CORLISS—*The Ciliated Protozoa*
Vol. 8. GEORGE—*The Brain as a Computer*
Vol. 9. ARTHUR—*Ticks and Disease*
Vol. 10. RAVEN—*Oogenesis*
Vol. 11. MANN—*Leeches (Hirudinea)*
Vol. 12. SLEIGH—*The Biology of Cilia and Flagella*
Vol. 13. PITELKA—*Electron-microscopic Structure of Protozoa*
Vol. 14. FINGERMAN—*The Control of Chromatophores*
Vol. 15. LAVERACK—*The Physiology of Earthworms*
Vol. 16. HADZI—*The Evolution of the Metazoa*
Vol. 17. CLEMENTS—*The Physiology of Mosquitoes*
Vol. 18. RAYMONT—*Plankton and Productivity in the Oceans*
Vol. 19. POTTS and PARRY—*Osmotic and Ionic Regulation in Animals*
Vol. 20. GLASGOW—*The Distribution and Abundance of Tsetse*
Vol. 21. PANTELOURIS—*The Common Liver Fluke*
Vol. 22. VANDEL—*Biospeleology—The Biology of Cavernicolous Animals*
Vol. 23. MUNDAY—*Studies in Comparative Biochemistry*
Vol. 24. ROBINSON—*Genetics of the Norway Rat*
Vol. 25. NEEDHAM—*The Uniqueness of Biological Materials*
Vol. 26. BACCI—*Sex Determination*
Vol. 27. JØRGENSEN—*Biology of Suspension Feeding*
Vol. 28. GABE—*Neurosecretion*
Vol. 29. APTER—*Cybernetics and Development*
Vol. 30. SHAROV—*Basic Anthropodan Stock*
Vol. 31. BENNETT—*The Aetiology of Compressed Air Intoxication and Inert Gas Narcosis*
Vol. 32. PANTELOURIS—*Introduction to Animal Physiology and Physiological Genetics*
Vol. 33. HAHN and KOLDOVSKY—*Utilization of Nutrients during Postnatal Development*
Vol. 34. TROSHIN—*The Cell and Environmental Temperature*
Vol. 35. DUNCAN—*The Molecular Properties and Evolution of Excitable Cells*
Vol. 36. JOHNSTON and ROOTS—*Nerve Membranes. A Study of the Biological and Chemical Aspects of Neuron-glia Relationships*
Vol. 37. THREADGOLD—*The Ultrastructure of the Animal Cell*
Vol. 38. GRIFFITHS—*Echidnas*
Vol. 39. RYBAK—*Principles of Zoophysiology—Vol 1*
Vol. 40. PURCHON—*The Biology of the Mollusca*
Vol. 41. MAUPIN—*Blood Platelets in Man and Animals—Vols 1–2*
Vol. 42. BRØNDSTED—*Planarian Regeneration*
Vol. 43. PHILLIS—*The Pharmacology of Synapses*
Vol. 44. ENGELMANN—*The Physiology of Insect Reproduction*
Vol. 45. ROBINSON—*Genetics for Cat Breeders*
Vol. 46. ROBINSON—*Lepidoptera Genetics*
Vol. 47. JENKIN—*Control of Growth and Metamorphosis*
Vol. 48. CULLEY—*The Pilchard: Biology and Exploitation*

PHYSIOLOGY OF ECHINODERMS

JOHN BINYON

Department of Zoology, Royal Holloway College,
Englefield Green, Egham

PERGAMON PRESS

OXFORD · NEW YORK · TORONTO
SYDNEY · BRAUNSCHWEIG

Pergamon Press Ltd., Headington Hill Hall, Oxford
Pergamon Press Inc., Maxwell House, Fairview Park, Elmsford, New York 10523
Pergamon of Canada Ltd., 207 Queen's Quay West, Toronto 1
Pergamon Press (Aust.) Pty. Ltd., 19a Boundary Street, Rushcutters Bay, N.S.W. 2011, Australia
Vieweg & Sohn GmbH, Burgplatz 1, Braunschweig

First edition 1972

Library of Congress Catalog Card No. 72-84199

Printed in Great Britain by A. BROWN & SONS LTD.

ISBN 0 08 016991 0

CONTENTS

LIST OF FIGURES

LIST OF TABLES

ACKNOWLEDGEMENTS

I AM grateful to the following holders of copyright for permission to reproduce various figures in this text: Academic Press Inc., for Fig. 1 from *Chemical Zoology*, vol. 3, Ed. M. Florkin and B. T. Scheer, p. 85, fig. 1; Fig. 9 from *Toxicol. Appl. Pharmacol.* **7**, 373-381, fig. 1; Fig. 13 from *Chemical Zoology*, vol. 3, p. 137, table 1; Fig. 16 from *J. Ultrastruct. Res.* **16** (1966), figs. 1 and 3; Fig. 24 from *Symp. Zool. Soc. Lond.* No. 20 (1967), p. 48, fig. 12. The American Association for the advancement of Science for Fig. 19 from *Science*, **145** (1964), p. 174, fig. 2; Fig. 20 from *Science* **138** (1962), p. 909, fig. 1; Fig. 21 from *Science* **146** (1964), 1178, fig. 3. The American Association for Limnology and Oceanography for Fig. 3 from *Limnol. Oceanogr.* **2** (1957), 192, figs. 13 and 14. W. H. Freeman & Co. for Fig. 17 from *Structure and Function in the Nervous Systems of Invertebrates*, by T. H. Bullock and G. A. Horridge (1965), vol. 2, p. 1527, fig. 26.5. MacMillan (Journals) Ltd., for Fig. 23 from *Nature Lond.* **201** (1964), 1344, fig. 1. The Council of the Marine Biological Association of the United Kingdom for Fig. 5 from *J. mar. biol. Ass. U.K.* **42** (1962), 52, fig. 1. Masson *et cie* for Fig. 2 from *Traite de Zoologie*, Tome XI, by L. Cuenot (1948), p. 146, fig. 174; for Fig. 18 from the same volume, p. 105, fig. 126. McGraw-Hill Book Co. (UK) Ltd. for Fig. 10 from *The Invertebrates*, vol. 4, by L. H. Hyman (1955), p. 143, fig. 58A. Pergamon Press Ltd. for Fig. 24 from *Comp. Biochem. Physiol.* **32** (1970), 748-9, figs. 1, 3, 4, 5. The Royal Society for Fig. 11 from *Phil. Trans. roy Soc.* **172** (1882), 829-885, figs. 23, 24, 25 and 26. Springer-Verlag for Fig. 14 from *Z. fur Zellforsch. mikrosk. Anat.* **60** (1963), 436, Abb. 5, and for Fig. 22 from *Z. Zellforsch. mikrosk. Anat.* **108** (1970), 457-474, figs. 2, 5 and 6. The Interscience Division of John Wiley Inc. for Fig. 8 from *Physiology of Echinodermata*, Ed. R. A. Boolootian (1966), p. 118, fig. 5-7; for Fig. 12 from the same book, p. 278, fig. 12-1, and for Fig. 15 from the same book, p. 221, fig. 9-1. The Wistar Press for Fig. 4 from *J. exp. Zool.* **140** (1959), 227, fig. 6.

E.J.B.

CHAPTER I

FEEDING AND DIGESTION

FUNDAMENTALLY the echinoderms are radially symmetrical animals and the overt expression, in all but a few species, of the pentamerous condition has contributed much to the uniqueness of this group amongst the invertebrate phyla. Such a symmetry has impressed either drastic modification or even absence of many of those systems commonly found in what may be thought of as more orthodox creatures.

The lack of a head or only a slight degree of cephalization has resulted in a body form where the mouth is often closely applied to the substratum or, conversely, faces vertically upwards. The position of the anus, if present, is variable and evolution has been frequently directed to the making of satisfactory sanitary arrangements. Probably as an indirect consequence, the nervous system is circumorally disposed with major extensions into the arms or ambulacra. A true circulatory system with the normal transport functions is lacking and recruitment for such requirements has been made on a rather *ad hoc* basis. The basic layout of such systems will be illustrated where appropriate in the text for the benefit of those not familiar with echinoderm anatomy. As species have become adapted to a variety of diets, so the digestive tracts and the enzyme complements have become similarly diversified and there is probably no better way to start a detailed consideration of the physiology of this group than with the digestive system.

Crinoidea

Although classically separated upon diverse morphological grounds, the subphylum Pelmatozoa would seem to form a trophic as well as a systematic grouping. Whilst exact details of the feeding process in

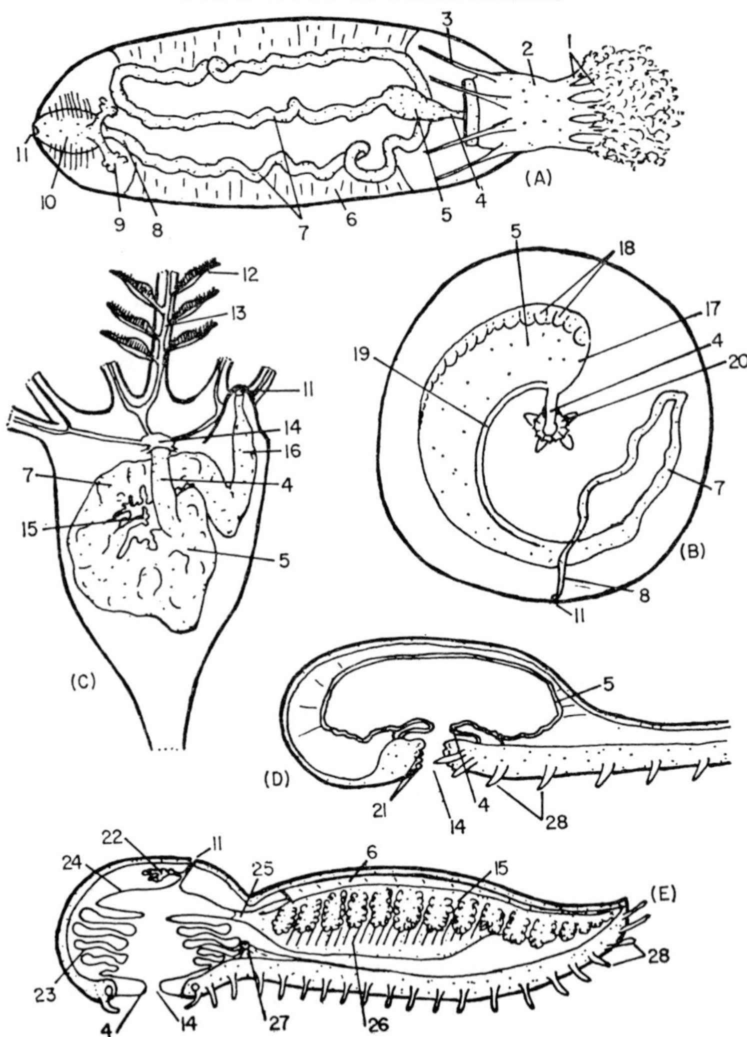

Fig. 1. Diagrammatic representations of the major features of the digestive systems of different types of echinoderms. A, sea cucumber; B, sand dollar; C, sea lily; D, brittlestar; E, starfish; 1, tentacles; 2, introvert; 3, retractor muscle; 4, oesophagus; 5, stomach; 6, mesentery; 7, intestine; 8, rectum; 9, respiratory tree; 10, cloaca; 11, anus; 12, pinnule; 13, food groove; 14, mouth; 15, digestive diverticulum; 16, anal sac; 17, caecum; 18, festoons; 19, siphon; 20, lantern; 21, teeth; 22, rectal caecum; 23, cardiac stomach; 24, pyloric stomach; 25, pyloric duct; 26, Tiedemann's pouch; 27, retractor harness; 28, tube feet.

fossil forms can only be surmized, it can be investigated in the still extant Crinoidea, for, whether they be sessile sea lilies or mobile feather stars, all crinoids feed in a basically similar fashion. These animals fix themselves down by the cirri to the substratum to feed, so as to expose the maximum length of ciliated ambulacral groove which is made sticky by the secretion of mucus. The efficiency of this groove, which is upwardly directed towards the main source of detritus, can be increased by the behavioural posture adopted by the hungry animal and its total length in some forms may be in excess of 100 yd.[137] The chief food of these animals consists of various protozoans, unicellular and filamentous algae, diatoms, larvae, small crustaceans and detritus generally. These are all flicked into the groove, which runs along the middle of each arm and its attendant pinnules, by the activities of the tube feet and are then conveyed to the mouth. The gut is coiled and ciliated for most of its length. It is longer in those forms with a reduced ambulacral system, such as the Comasteridae; this family, however, may use the terminal combs of the oral pinnules as accessory feeding structures.[303] The suggestion that the mucus secreted contains a toxic principle has not been confirmed.[658]

Nothing is known of the process of digestion itself in this group, even to the extent of whether it is entirely intracellular or if enzymes are liberated into the lumen of the gut. This represents a gap in our knowledge of the comparative physiology of what must surely be the primitive feeding mechanism of this whole phylum. The ciliation of the disc often seems rather ineffective in removing the faeces which are voided from the anal papilla.

Holothuroidea

In the eleutherozoan groups, only the Holothuroidea are still wholly microphagous: in the remainder this simple form of feeding has, at least in part, been replaced or supplemented by more complex mechanisms which permit a wider range of diet and, consequently, of habitat. The Holothuroidea, which form a conspicuous element of many littoral faunas, fall broadly into two main trophic groups. Firstly, the dendrochirotes, which extend their mucus-covered feathery tentacles into the sea water just above the sand where they live. Minute organisms of various types adhere to them and then, one by one, the tentacles are "sucked" clean by insertion in the mouth. In addition the animal may "sweep" an area of sand ahead of it for detritus. Secondly, the aspidochirotes use their peltate tentacles as shovels, ingesting large quantities of sand, whilst the burrowers just engulf the substratum as they move through it, rather as does the common earthworm in the soil. Specialized pelagic forms, such as the elasipod *Pelagothuria*, feed upon plankton. Feeding may be continuous or intermittent and may be restricted to the hours of darkness in those species where a pronounced diurnal rhythm exists.[821] Food, consisting chiefly of protozoans, diatoms, algae and larvae as well as detritus, passes through the gut largely by peristalsis which can persist in isolated preparations for some time.[604,704] The period taken for the passage of food has been variously estimated at from 3 to 36 hr.[604,768,749,816] Where the holothurian population is fairly dense, the quantity of sand ingested per annum may be quite appreciable; Yamanouchi[815] gives a figure of about 58 kg/year/animal. According to Crozier[170] and other workers, between 10 and 60 tons/acre will be passed through the guts of holothurians. This compares with Darwin's classical figure of 10 tons/acre for earthworms, although more recent estimates, which allow for that soil which does not appear as casts, put the figure at between 4 and 36 tons/acre.[224] Bonham and Held[72] have demonstrated that on one Pacific atoll approximately $2{\cdot}4 \times 10^8$ kg of sand per annum was being passed through the guts of the holothurian population. It would seem that the holothurians materially assist the annelid population of the sea bed in performing a similar function to that of the earthworm population on land. However, unlike earthworms, particle comminution has not been detected, but their stomach pH of around 5·0 would be sufficient to dissolve some calcareous material. The digestive tract often contains a yellowish fluid, the aroma of which results from the presence of unsaturated fatty acids,[705] and tends to disappear after a period of starvation. The pH of the gut fluid varies during the process of digestion and is generally more alkaline towards the large intestine. Numerous enzymes have been reported both from the digestive juice and from extracts of the gut wall, including a protease, invertase, lipase but no amylase from *Thyone briareus*.[782] A similar battery of enzymes has been reported in *Holothuria*[606b] including amylase and maltase. In his studies upon *Caudina chilensis*, Sawano[691] lists a tryptic protease, maltase, invertase, glycogenase and lipase. A range of carbohydrases including amylase, cellulase, pectinase, dipeptidase and a weak lipase have been described in *Stichopus japonicus*.[135] Franssen and Jeuniaux[268] suggest that alginase is a likely constituent of the digestive juice of *H. forskali*.

In a review of digestion in *H. atra*, Trefz[768] has shown the presence of a strong free lipase in the gut. Trefz also questions the long-held concept of the transport of enzymes into the gut by amoebocytes. The enzymes were originally thought to be stored in the gut wall as yellow granules. These granules have subsequently been shown to be lipoidal in nature which makes their identity as enzymes unlikely. The presence of free lipase in a specimen with a regenerating gut, and also in *Thyone*, neither of which possesses a rete mirable, suggests a more conventional mode of enzyme secretion.

In addition, the fact that there is no increase in fat globules after feeding the animals with milk, coupled with the observed insufficiency of amoebocytes (thought to be migrating from the gut lumen towards the coelom anyway), makes such secretion, at least for lipase, seem more probable. The yellow granules may be functionally related to the yellow fluid already referred to.

Amoebocytes have, however, been shown to be involved in phagocytic ingestion in the gut. Iron saccharate is, for example, taken up in this way, passes into the haemal system and is eventually eliminated by way of the respiratory trees, chiefly the left branch.[741,768] The left branch seems to possess more red blood cells and is more obviously respiratory.[741] Recently Fish[244,245] has demonstrated the presence of a tissue fluid system beneath the mucosal epithelium which is continuous with the haemal fluid. This fluid contains amoebocytes, a number of amino acids and is PAS positive. Histochemical tests suggest that the cells bordering the lumen of the intestine elaborate enzymes which include trypsin-like endopeptidases, several exopeptidases, maltase, invertase and esterases, lipid appearing to be the chief storage substance. Interestingly, Tanaka[749] reports that the nitrogen content of the food in the gullet of *S. japonicus* is some three to five times that of the bottom material. He also showed that the sugar content of the perivisceral fluid fell from 6·4 mg% in the field to only 1·1 mg% after 165 hr starvation. Seasonal feeding would seem to occur in this species as it does in *Cucumaria elongata*.[246]

The holothurian gut is, however, still something of an enigma. It appears to act as an effective barrier to chloride, sulphate, urea, glucose, pentose, saccharose and the dyes methylene and trypan blue.[342,606a,704] This led Enriques[220] to propose the idea that the passage of enzymes and digestion products through the gut wall occurred by means of amoebocytes, a theory which found general acceptance for many years. Even as recently as 1965 Hetzel[346] suggested that the morula cells play an important role in amino acid uptake from the digestive tract, and were also involved in protein synthesis and storage. He believed these cells originated in the haemal vessels and not in the perivisceral coelom as was thought by Endean.[217] The passage of enzymes into the gut has already been questioned and recently the permeability of the gut wall to the products of digestion has been reinvestigated by several workers.[182,452,682] The gut of *Leptosynapta inhaerens* appears very permeable to dextrose, transporting it at a rate of some 16 μg/min. This rate was markedly reduced by magnesium chloride, suggesting that simple diffusion alone was not responsible and that some form of active transport was involved. Other experiments recorded a value of 6 μg/hr/mg dry weight and in this instance transport was inhibited by phlorizin. However, the gut of the related holothurian *S. parvimensis* did not transport monosaccharides and simple diffusion cannot be ruled out in all cases. Some amino acids seem to be transported across the intestinal wall, but no movement of sodium or chloride ions was detected and no potential difference was developed across it. These results do not rule out the possibility of solute transport across other membranes such as the respiratory trees or body wall. Suspended organic materials such as fat droplets or amoebocytes loaded with yeast cells have been shown, for example, to enter the coelomic cavity by way of the respiratory trees. It is doubtful, however, if this is really an accessory feeding mechanism and the whole problem of feeding, digestion and transport of nutrients could well bear reinvestigation using modern techniques. Similarly the part played by the haemal system is not well understood, especially since it is in this group of echinoderms that it reaches its fullest development with a true circulation of fluid within a capillary network[390] resulting at least in part from the contraction of the main dorsal vessel.[644]

Echinoidea

The Echinoidea or sea urchins are the only group of living echinoderms still to be considered which lack oral arms and which must, like the holothurians, make other provision for obtaining their food. The regular echinoids possess a complex device called Aristotle's lantern which is used for this purpose. The structure is well described in standard zoological texts, but briefly consists of a pentagonal frame-

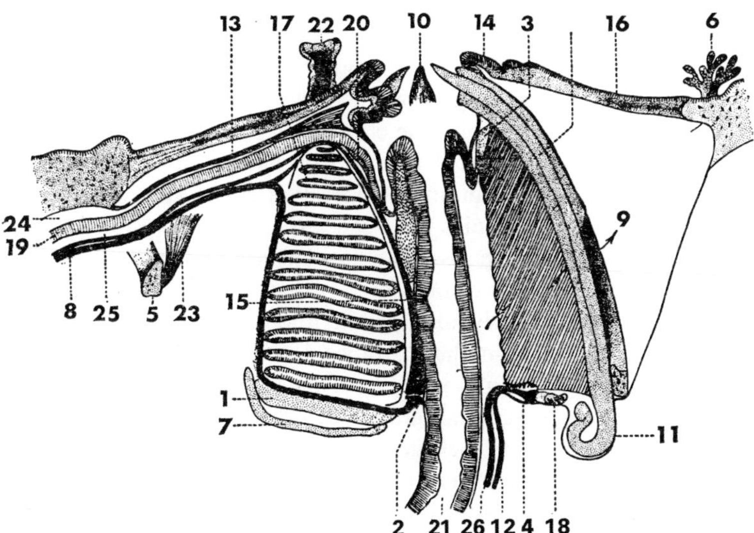

FIG. 2. Sagittal section of the chewing apparatus and oral surface of *Echinus esculentus* passing through the A radius and CD interradius. 1, rotulus; 2, circumoral water vessel; 3, nerve ring; 4, spongy body; 5, auricle; 6, peristomial gills; 7, compass; 8, radial water vessel; 9, peripharyngeal cavity; 10, tooth; 11, tooth sac; 12, haemal channel to axial organ; 13, ligamentous membrane; 14, buccal lip; 15, radial haemal sinus; 16, peristomial membrane; 17, interpyramidal muscle; 18, rotular muscle; 19, radial nerve cord; 20, deep oral nerve mass; 21, oesophagus; 22, buccal podium; 23, retractor muscle; 24, epineural sinus; 25, hyponeural sinus; 26, stone canal.

work of ossicles supporting five vertically arranged and continuously growing teeth. These teeth, which are made of an especially hard calcite, can be variously moved by the musculature. This movement can be supplemented by movement of the lantern as a whole by protractor and retractor muscles attached to the test, although this process is thought to be more respiratory and locomotory than masticatory. A certain degree of rotational movement can be imparted to the structure and additionally the teeth can be ground one upon another for triturating food material by the comminator and other muscles. (For recent description, see Cobb and Laverack.[144])

Primarily, the sea urchin is a browser and scavenger, the chief food being encrusting algae and seaweeds, while the more carnivorous ones take barnacles, hydroids, tube worms, sponges, ascidians and occasionally dead fish and other carrion. The Echinothuridae feed upon ooze as do other deep water echinoids.[464] This diet is supplemented by smaller organisms caught by the activities of the pedicellariae. Although the physiology of these effector organs will be considered in more detail later on, suffice it here to say that the globiferous variety have a toxic bite which can immobilize small crustaceans and the like. The feeding rate in echinoids has been less intensively studied than in holothurians, but nevertheless they do constitute an important factor in the bionomics of the littoral zone. Forster[260] has shown that the *Echinus esculentus* off Plymouth graze about 2 cm^2 of rock surface per hour. This results in the clearance of about one-third of all the available rock surface in a year. This must surely have considerable effect upon the settlement and growth of other marine organisms. The bionomic importance of such grazing should not be underestimated. Schaefer,[697] for instance, has observed that the urchins *Strongylocentrotus purpuratus* and *S. franciscanus* have a devastating effect upon kelp beds and these animals would move onto such beds from some little distance. Marine grasses too are subject to attack by species of *Lytechinus*, *Tripneustes* and *Diadema*.[650] Surprisingly, this latter species has been shown to have a "taste preference" for certain other echinoderms when hungry,[648] although its growth rate was maintained upon plant food.[651] The rate of growth of *Diadema* is, nevertheless, low when compared with other tropical echinoids.[466] Field observations upon growth rates within a population can sometimes be misleading as size is not always a good indication of age and negative growth rates have been recorded under laboratory conditions and in the field.[215] This probably occurs as a result of resorption of the peristomial plates as was described by Cutress.[180] The assimilation rate has further been shown to vary with the animals diet by Fuji.[282] When *Laminaria* was was the only alga available, *S. intermedius* consumed about 3 g wet weight/day, but when *Ulva* was solely available only 0·5 g was taken. He also demonstrated a seasonal decline in the feeding rate between July and October and also a reduction in the percentage of nutrient extracted from the food material from 70% in summer to 55% in winter. *Tripneustes* and *Lytechinus* too have a seasonal feeding rate which is adapted to the ambient temperature but the assimilation efficiency is fairly constant at around 50-60%, due to the higher respiratory rate in summer.[551] On a *Thallasia* community producing 8·8 kg/m^2/year, the grazing of these urchins accounted for between 0·7 and 5·5 kg/m^2/year, a not inconsiderable percentage of the total grass production. The large decapod crustaceans in Lough Ine had a sparing effect upon the algal population by devouring the urchin *Paracentrotus lividus*.[558]

The spatangoids and clypeastroids feed in an entirely different manner from the regular urchins. These species normally live in sandy burrows and lack the lantern apparatus, although a degenerate form is to be found in the latter group. Not surprisingly, therefore, their gut contents have been shown to contain quantities of bottom material largely sand with diatoms, foraminiferans together with fragments of larger animals including worms, molluscs, coelenterates.[303] Mucus strings with embedded food particles are brought to the mouth by ciliary currents.[565] Goodbody,[307] working with *Mellita sexiesperforata*, showed that sand falling on to the aboral surface as a consequence of burrowing is sorted out by club-shaped spines. Smaller particles drop down between the spines and are carried to the ambulacral grooves and thence to the mouth. Feeding stops during defaecation. Total dry weight production by the irregular urchin *Moira atropos* was shown to contribute as much as one metric ton for every tenth square mile of sea bed.[550]

Food entering the gut of urchins takes about 4 hr to reach the intestine and may require up to 2 weeks before the undigested remains are voided. Earlier workers have shown the presence in the stomach of a protease, amylase, a weak invertase, but no lipase, the gut being slightly more acid than sea water. In spatangoids an amylase and a protease in acid secretion are said to be present in the large caecum. More recent work[359,450] has demonstrated the additional enzyme iridophycase as well as the presence of an agar digesting bacterial flora in the second loop of the intestine with a density of some 10^6/ml. In isolated culture, these bacteria do not, however, liberate glucose into the medium so the analogy with terrestrial ruminants cannot be too closely drawn. Interestingly, a number of species of yeast also have been isolated from sea urchins.[181] Generally speaking, these animals possess a good battery of carbohydrases of one sort and another, for dealing with their varied diet. Some 76% of the dry weight of the alga *Macrocystis pyrifera* can be assimilated by the urchin *S. purpuratus*. As this alga possesses some 36% of its dry weight in the form of algin, it is mandatory that either an alginase or some form of bacterial decomposition should exist in this animal.[440] An algin depolymerase has in fact also been recorded,[221] and alginase in *Psammechinus miliaris*.[268] The presence of cellulase has been reported in three species of echinoid,[830] and α- and β-glucosidases as well as proteases and weak lipase are reported being available for digestion in *D. antillarum*.[465] This animal incidentally ingests much calcium carbonate with its food and is capable of eroding coral reefs.

Although enzyme secretion and the transport of nutrients is traditionally the function ascribed to amoebocytes in echinoids, Farmanfarmaian and Phillips[232] thought that the demonstration of phagocytosis and the suspected migration by these cells to be poor evidence for such transport. They fed the purple sea urchin *S. purpuratus* with a radioactive alga. Activity appeared chiefly in the oesophagus and stomach, that in the perivisceral fluid reaching a peak in about 24 hr and then falling to a constant level for many days. It was only during this "levelling off" period that the coelomycetes became strongly labelled. Some 90% of the activity in this fluid was due to galactose derived from galactosylglycerol present in the alga. Although the first four festoons of the intestine were free from bacterial enrichment, there was some enrichment farther along the gut. The addition of antibiotics, however, did not influence the rate of appearance of radioactivity in the coelomic fluid. Severence of the sinus connecting the aboral perihaemal ring to the gonad did not influence the build up of radioactivity in that organ, suggesting that the haemal system, albeit a fairly extensive one, plays little or no part in the transport of nutrients and its function is still something of a mystery.

Boolootian and Lasker[85] showed that it was the red haemocytes, laden with glycogen in well-fed animals, which transported nutrients, for the injection of radioactive haemocytes into an inactive recipient led to the distribution of activity to all tissues. These authors note the numerous connections between the dorsal marginal sinus of the second festoon of the intestine and the gonad, and think that this route may account for the transport of a significant proportion of nutrients to the gonad. Burton[105] has been unable to confirm the presence of glycogen in similar red granule containing amoebocytes of *D. antillarum* and *P. miliaris*. She did confirm the presence of albuminoid inclusions and also of amoebocytes with iron containing inclusions, which makes the results obtained from feeding experiments using iron saccharate somewhat unreliable. This author was unable to present any conclusive evidence of cellular participation in nutrient transport.

Amino acids have been shown to appear in the perivisceral fluid of *Arbacia* following protein digestion.[782] The concentration of reducing sugar rises after a meal rich in carbohydrate and the concentration of non-protein nitrogen rises after a meal rich in protein and there is a rapid disappearance of glucose injected into the coelom of a starving sea urchin.[450] Cyclical changes in these substances are associated either with feeding rates or the reproductive state, and have been studied in several sea urchins.[46,300,313] Field observations indicate a seasonal change in the protein concentration of the coelomic fluid,[352] although it never falls as low as it does in animals starved under laboratory conditions. The presence of carbohydrate, both free and bound as glycoprotein has been demonstrated in *A. lixula*[506] as also has a range of amino acids.[302]

The circulation of the coelomic fluid must obviously play an important part in the translocation of nutrients in these animals. The gut as well as the gonad function as storage organs, for echinoids lack any form of gut caeca and many types of inclusions and droplets have been described. These include glycogen, which nearly disappears from the gonad just prior to spawning, and is especially noticeable in the antarctic echinoid *Sterechinus neumayeri* which had discrete reproductive periods,[620] lipids, alkaline phosphatase[281] and sulphated acid mucopolysaccharide.[353] Glycogen is also abundant in the muscles[45] and in the last half of the stomach and first half of the intestine. This too decreases during starvation.[78,453] The true function of the siphon too is still uncertain although it has been the subject of a number of investigations. Waves of contraction passing along it, together with a current of water,[341] have been confirmed by Stott[740] who also suggested a possible respiratory function, although Cuenot[179] thought that it acted as a by-pass for the excess water taken in with the food, thus allowing a greater concentration of enzymes to be reached in the stomach.

An unusual form of nutrition in both regular and irregular echinoids has recently been described by Pequignat.[622] He records the flow pattern of particles taken into the burrow of *Echinocardium cordatum* and notes that such particles are often deflected onto a "limy mucus cap" covering the flat dorsal spines. In the periproctal region of this cap he claims that extracellular digestion and absorption occur. Using coagulated albumin stained with acid fuchsin and gelatin flakes soaked with ninhydrin he demonstrated an increasing coloration of the underlying epidermis when these were placed upon the dorsal furrow. Colouring matter is only released from these preparations upon hydrolysis and the coloration of the integument due to the localized immigration of coloured corpuscles was distinguishable from that due to straight staining with acid fuchsin. Eventually even the connective tissue fibres became reddened due to transport of the dye by amoebocytes, which lends support to idea of Liebmann[468] regarding the function of trephocytes, none of which were seen returning. It was possible, by incorporating synthetic peptides in the gelatin, to characterize the presence of a leucine aminopeptidase. Polysaccharides and fat are similarly transported and *P. miliaris* was even observed to digest small animals such as *Natica catina*. These observations were extended to *Asterias rubens* and he suggests that the "secretory packets" observed by Chaet and Phillpot[124] may be enzymic in nature. *Ophiothrix fragilis* also exhibited brachial digestion and absorption and he offers a similar suggestion as to the nature of Buchanan's A and B bodies.[93] An interesting correlation is made between the possession of this type of nutrition and the organization of the echinoderm gut and also the lack of exoparasites, superficial ciliates and bacteria. Significant amounts of soluble organic material are to be found in the seawater in which *Strongylocentrotus droebachiensis* lives, but this has not so far been demonstrated to be a food source.[401]

Asteroidea

Both the Asteroidea and the Ophiuroidea possess oral arms with ambulacral grooves lined with serially arranged tube feet. Consequently it might be expected that their feeding mechanisms would more closely resemble those of the Crinoidea. However, although a few asteroids still retain a primitive ciliary-mucus feeding mechanism, the majority now employ their tube feet, often with well-developed suckers, for a more active carnivorous mode of feeding. In *Porania*, a microphagous feeder, the detritus falling upon the aboral surface is conducted to the mouth by way of ciliated grooves, as it is in *Ceramaster placenta*.[793] Possibly the Ordovician starfish *Villebrunaster* may have fed in a similar way. In *Ctenodiscus* the chief source of food is the mud in which the animal lives. Gislen[303] thought that the cribriform organs, which are vertical depressions on the sides of the arms carrying ciliated lamellae, sieved off the coarser particles. It seems likely that several species, such as *Luidia* and *Astropecten*, use ciliary-mucus feeding as an adjunct to their more usual method.

The distensible mouth and the eversible stomach must have contributed as much to the success of the asteroids as carnivores as has the elaboration of the water vascular system, for many asteroids, particularly phanerozones, swallow their prey whole. This category includes such animals as the above-mentioned *Astropecten*, which additionally feeds upon small gastropods and bivalves, together with other echinoderms and crustaceans.[360,807] Similarly *Luidia*, which seems to prefer foraminiferans, and many other species may have a particular and well-defined dietary preference. *Solaster* is well known as a predator upon *Asterias* and again *Luidia* will also devour the ophiuroid *Ophiura albida*.[239] Even on this rather calcareous diet a conversion efficiency of 33% is claimed. It is even within the capabilities of the wafer-thin *Anseropoda* to catch and eat small crabs, and some of the larger starfish manage to swallow sea urchins and in most cases the undigested shells and tests are passed out through the mouth.

The use of the eversible stomach can be best seen in those asteroids which deal with the larger bivalves, such as *Asterias*, although even in this genus the diet can vary considerably.[318,319,320,234b] The mechanism of stomach eversion is dealt with in Chapter 11. The exact method by which the adductor muscle of their prey is overpowered has been the subject of a prolonged controversy. Additionally, it has brought the echinoderms to a position of economic importance, for one of the larger bivalves devoured by certain asteroids is the oyster. A number of control measures have been assessed,[284,318,461] chiefly in the U.S.A., but without great success. Although the loss sustained by the shellfish industry may be considerable, a further and less obvious effect of starfish is that as bottom feeders they may compete with other economically important species for food. Hatanaka and Kosaka[332] have calculated that starfish may take as much as 80% of those bottom-living animals which are taken as food by all species. Thus they are competing with plaice and other species valuable to man as food. They further show that a 2-year-old *A. amurensis* will have devoured some 1200 brittle stars. Recently much alarm has resulted from the destructive powers of large numbers of *Acanthaster planci* in the Pacific Ocean upon coral reefs.[134b]

During feeding the starfish humps itself up over the mollusc, which, if not attached to the substratum, is manoeuvred by the tube feet so that the edges of the shell are directed towards the mouth. The podia of the proximal regions of the arms attach themselves to the shells, whilst those of the more distal parts are attached to the substratum so as to give the animal a purchase. When in this attitude, Schiemenz[700] found that a force of some 1350 g was necessary to remove the mollusc. He further found that a force of only 900 g would cause the rugid *Venus* to gape in from 5 to 25 min. A continuously exerted force of 1000 to 2000 g would eventually lead to rupture of the adductor muscle. Following this purely mechanical explanation, it is not surprising that a pharmacological one should have been proposed. Stomach extracts of *Asterias rubens* were shown to have a toxic effect upon the frog sciatic/gastrocnemius preparation,[782] and similar extracts of three out of five species of Japanese asteroid had a toxic effect upon

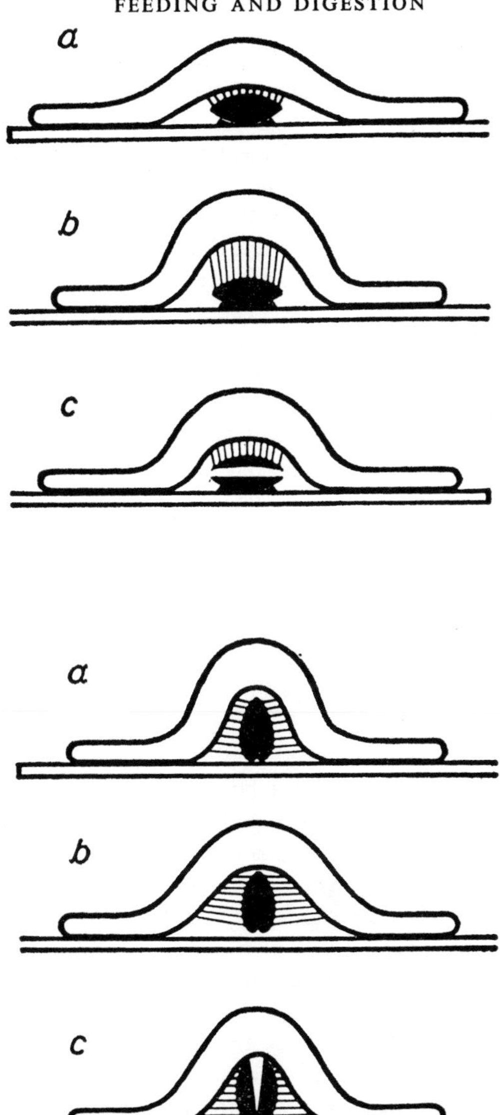

FIG. 3. Schematic drawings of stages of the opening of an attached clam by *Evasterias*. Lower set indicate opening of an unattached clam by the same starfish.

the heart of oysters.[693] It was of special interest that those species containing a toxic principle were those which normally fed upon oysters.

In 1942 Reese[657] showed that Schiemenz's values for the strength of the adductor muscles were on the low side and consequently the mechanistic view lost some support. However, the only further evidence to be published concerning the possible toxic nature of asteroid stomach extracts seems to have been by Nakazawa in 1960,[571] who showed that extracts from *A. amurensis* caused cellular degeneration and disintegration of small coelenterates and blastulae. Nematocysts were also discharged and in these respects the extract acted like a calcium free medium. Here, so far as the existence of a toxin in the stomach is concerned, the matter rests. As will be shown below, the combined muscular and digestive powers of the starfish are sufficient for their purpose, but such evidence does not rule out the possibility of the existence of a toxin, especially in view of their widespread occurrence in other tissues throughout the phylum. Significantly, it is reported that "comet" forms are able to open molluscs.

The role of the tube feet and the forces exerted during the opening of bivalves was reinvestigated in the 1950s by a number of workers,[101,136,234a,451] and it was during the course of this work that a significant new fact came to light. Using *Pisaster ochraceous*, Feder[234a] showed that this starfish employed feeding techniques which depended upon the size of the prey. With a large *Mytilus galloprovincialis* the starfish insinuated its stomach through the byssus opening and digested the mussel inside without even attempting to draw the valves apart. He then bound some *Mytilus* with wire and the result was still the same, even when the starfish possessed fewer than three arms. The measured width of the byssus opening was of the order of 0·2 mm. On smaller mussels, a starfish of some 30 cm diameter pulled the valves apart, exerting a force of between 1200 and 2600 g. In many cases the animal humped itself up over the region of the posterior adductor muscle. It was further pointed out that the opening of clams may be facilitated by their own marked insensitivity to the presence of a starfish, even though the starfish may actually be digesting the clams tissues. This is in contrast with Bullock who in 1953[98] demonstrated that even the presence of a tube foot upon the mantle edge of a number of molluscs including clams, released a very definite sequence of escape responses. For example, the limpet *Acmaea* will move rapidly away from the starfish *P. ochraceous*, showing no tendency to clamp down and may even swirl the shell about so that the starfish cannot get a grip. *Acmaea* does not react in this way to echinoderms which do not normally prey upon it, suggesting a chemical difference between species (see Chapter 6).

Lavoie[451] using prepared clams in which the adductor muscles were replaced by springs also demonstrated forces of up to 3 kg and underlined the important point mentioned by Feder, namely that a gape of only 100μ was necessary for the insinuation of the stomach, a gape which is not much more than twice the thickness of the stomach wall itself. Solvent extracts of the stomach affected the heartbeat of the clam, but so also did sea water.

Experiments by Burnett[101] showed that a mussel would be completely digested in 24 hr when a large hole was cut in the shell and then sealed over with fine bolting silk through which the stomach could not pass. As the heart of the mollusc was still beating after 5 hr it suggests a purely digestive process and that no toxin was liberated, at least in the early stages.

In view of the long periods for which the valves can be kept shut and the low metabolic activity of the contractile mechanism of the adductor muscle, it seems reasonable to suppose that a copious supply of blood is not essential. It is to be wondered, therefore, whether the action of the supposed toxin should be sought, not upon the heart, but upon the adductor muscle. Christensen[136] recorded even greater forces of up to 5·5 kg and comments upon the ability of the starfish to "lock" itself in position before beginning to feed. There exists, therefore, the intriguing situation of a physiological "arms race", the predator having evolved some kind of locking mechanism of its own in order to overcome the more familiar "molecular catch" mechanism of the adductor muscle of the prey. It has been observed by Hancock[321] that the feeding rate of *Asterias rubens* was higher upon *M. edulis* from British waters than upon those mussels from Danish waters. The Danish mussels possessed the larger and the more stringy adductor muscles.

Some asteroids exhibit feeding rhythms, for in a population of

P. ochraceous only about 5% were feeding at any one time during January and February,[501] but some 60 to 80% were feeding in the summer months. Ingestion rates have also been found to vary with season, ranging from 3 g dry weight/tidal cycle/100 animals in winter to ten times this figure in summer.

During the feeding process the main enzyme secreted by the stomach wall is a proteinase which is, apparently, tryptic in nature. Emulsifying agents are also said to be secreted to assist in the disintegration of the food material. The resulting soup is conveyed into the animal partly by retraction of the stomach, but principally by the flagellated gutters which lead in some instances right up into the pyloric caeca.[10,11,12,13,14] These may also serve as rejection channels. Passage of food through the gut is by ciliary action rather than by peristalsis, the caeca being particularly well endowed in this respect. Additionally an amylase has been reported,[133,738] and later denied[782] from extracts of the stomach wall, but invertase is certainly present. Extracts of the pyloric caeca have been variously shown to contain protease, amylase, and a lipase by the above authors, operating in a medium more acid than sea water.[663] The time taken to completely digest a piece of *Pecten* muscle was found to be about 60 hr. The diffusion through the wall of the isolated gut caeca of *Paturia miniata* of the products of digestion of gelatin has also been demonstrated. However, as this species possesses an extensive Tiedemann's diverticulum underlying the caeca, these experiments may be criticized on the grounds that digestion could have occurred in this region and not in the caecum itself. A variety of proteinases has been found in caecal extracts, including the intracellular cathepsin from both the pyloric and rectal caeca.[692] Ammonium carminate and iron saccharate injected into the stomach are taken up by the cells lining the caeca and fats seem to be ingested in the same manner, the lipolytic enzymes being intracellular. Thus in the more carnivorous asteroids the enzyme system is biased towards the proteolytic components compared with the essentially herbivorous and omnivorous holothurians and echinoids where the carbohydrases predominate. Asteroids are not, however, completely lacking in such enzymes for β-polyglucosidases are reported from *P. miniata*[20] and also that lichenin and laminarin are readily hydrolysed.

Much of the epithelium of the gut is devoted to food storage, which takes the form of glycogen, fat and a small amount of proteinaceous material.[10] The size of the caeca and gonads vary inversely with one another in the quantity of stored material and cyclical changes are to be observed.[231,313,501] The caeca are generally rich in lipid, containing some 10% of their wet weight of fat. Evidence seems to support the conclusion that *de novo* synthesis of fat occurs in the caeca of the starfish *Pisaster ochraceous*. The rate of incorporation of C^{14} labelled acetate into lipid is dependent upon the nutritive state and respiration rate, it being twice as high immediately after feeding than after 1 week's starvation. It was also maximal at the time of gametogenesis during November to January. Glucose is a less efficient precursor of lipid, but a more efficient precursor of polysaccharide. The idea of synthesis is further supported by the distribution of the polar lipids, largely phosphatides. The body wall and testis contain some 70% of their total lipid in this form, in the ovary 55% and only 30% in the caeca. The higher proportion of neutral fat in the caeca indicates the storage form, the caeca of *A. forbesi* containing three separate categories of glycerol-containing lipids.[216] The sterol fraction will be considered in Chapter 5. That the body wall may also function as a storage organ, is indicated by analysis of the Antarctic asteroid *Perknaster fuscus antarcticus*.[620,621] This species has a body protein content of 47% compared with only 12% in *Patiria miniata* and 14% in *Pisaster ochraceous*.

The haemal system being much less extensive than in the holothurians is less amenable to experiment. The part it plays in the distribution of nutrients is not clear, but glucose and amino acids injected into the coelom are rapidly taken up by the coelomocytes.[782] More recently the rapid uptake by all tissues, especially the pyloric caeca and coelomocytes of C^{14} labelled algal protein hydrolysate, glycine and glucose from the coelomic fluid[240,241] has been demonstrated. A rather high proportion (up to 50%) of the total nitrogen of the coelomic fluid is in the form of nutrients. However the non-protein nitrogen concentration of the perivisceral fluid of *P. ochraceous* seems to be a constant proportion of the total nitrogen content over different phases of the reproductive cycle.[785] Prolonged starvation had a more marked effect upon the protein concentration than upon the non-protein nitrogen component. Brief starvation, interestingly, slightly increases the NPN fraction followed by a loss of protein which is greater than can be accounted for in the elevated NPN fraction. The suggestion is that tissue nitrogen is mobilized, the protein content of the perivisceral fluid being used as an energy source. Transport is most likely to be accomplished by a continual flux between the various tissues and the perivisceral fluid rather than by a solely distributive activity on the part of the haemal system. Such flux would seem to be inhibited by metabolic inhibitors such as iodacetate. The turnover rates of this system are high, as much as 0·5 g of protein and 12 mg of glucose per day in an *in vitro* preparation of the gut caeca. These values differ markedly for related compounds with the suggestion of competitive inhibition.

The ability to absorb nutrients direct from the external medium has often been suggested as an additional source of food material. In echinoids, it will be recalled that solid matter could be digested when in contact with the epidermis, presumably by some form of extra-cellular enzyme secretion. Whilst such digestive processes have yet to be described for asteroids, absorption of exogenous nutrients is quite well documented.[735] Two species of asteroid, *A. forbesi* and *Henricia sanguinolenta*, are capable of absorbing glycine from dilute solutions in sea water.[735] Further experiments with *A. forbesi* demonstrated the appearance of significant amounts of radioactivity in the region of the perihaemal sinus and along the luminal borders of the tube feet when these animals were fed with C^{14} labelled glycine and glucose.[240a] Evidence suggested that nutrients may be absorbed directly via the epidermal cells as a supplement to the more orthodox mode of nutrition. In similar experiments the everted stomachs of *Patiria miniata* were shown to absorb protein from solution, although the concentrations used were higher than would normally occur in sea water.[20] Exposing these same two species of asteroid to low dilutions of C^{14} labelled amino acids and glucose, resulted in a large proportion being taken up by the external tissues, especially by the aboral surface, probably the tube feet.[243] Little of the total radioactivity so absorbed was lost from the animal after three weeks and there was no significant movement into the deeper tissues. A progressive decrease in the alcohol solubility of the radioactivity of the amino acid component suggests incorporation into proteins and a similar loss from the glucose fraction suggests its use as an energy source. The higher concentration of nutrients in the sea water sometimes resulted in the inflation of *Henricia* with the subsequent absorption of glucose by the digestive tract. The lower concentrations failed to elicit such a response. No difference in the rate of glycine uptake between normal and stretched gut caeca of *Echinaster spinulosa* could be found and it was concluded that the number of absorptive sites therefore remained constant. These caeca, *in situ*, underwent changes of tonus and such changes although affected by various drugs were unaffected by various food substances such as amino acids and monosaccharides, and were most likely an adaptation to aid the circulation of coelomic fluid.

Ophiuroidea

Although still to a large extent microphagous, the Ophiuroidea do not employ the orthodox variant of the ciliary/mucus feeding mechanism, for the simple reason that the appropriate ciliation is absent. The bottom material accounts for a large part of the diet, containing as it does, dinoflagellates, diatoms and foraminiferans. Amphiurids burrow leaving just the tips of the arms above the surface, which sweep about for detritus assisted by currents derived from the pumping action of the disc. Kelp fronds and hydroids may serve as food for some species, and polychaetes, small crustaceans and occasionally small bivalves are also taken. The guts of *Ophiothrix* and *Ophiopholis* have been shown to become radioactive when kept in water with C^{14} labelled phytoplankton.[678] The spines and tube feet of ophiuroids are well supplied with mucus glands[94] and it is not clear whether such internal radioactivity was a result of direct assimilation from the medium or by ingestion of contaminated mucus. *Ophiocomina*, however, does not assimilate material in this way, although it has been shown that it produced boluses of mucus which were passed to the mouth by the tube feet.[259] This species feeds in several different ways including the use of the tube feet for passing small animals directly to the mouth, rasping carrion,[570] and the employment of the coiled tips of the arms for ensnaring small prey, whence it is passed to the mouth.[570,792] This latter method is reminiscent of that employed by the deep sea gorgonocephalids, where subdivision of the arms has greatly increased the efficiency of the mechanism. In *Astrophyton muricatum* where batteries of hooklets are orientated into the direction of the current, the food consisted of some 85% clupeid larvae, 10% stomatopod larvae and 5% copepods.[190] The arms of the shallow water *Astrobia nuda* are also extended in feeding.[769]

The digestive physiology of the ophiuroids is practically unknown, a strong proteinase acting in both acid and alkaline media, an amylase and possibly a lipase being reported.[362] The haemal system has the typical echinoderm distribution, but nothing is known of its function. The direct absorption of nutrients from sea water has also received confirmation in this group of echinoderms[736] and *Ophiactis simplex* can accumulate up to 35 times the concentration in the external medium of C^{14} labelled glycine where a level of $2{\cdot}6 \times 10^{-8}$M was maintained.[736] A similar effect was observed using alanine, arginine and valine. At higher external concentrations the degree of accumulation tails off and this may account for the negative results of some authors. It was also observed that at reduced salinity a higher proportion of the absorbed radioactivity was to be found in the alcohol-insoluble fraction of whole body extracts.

In summary it may be said that the quantity of food actually assimilated is small (except possibly the carnivorous asteroids), although the bulk of material ingested may be quite large. This is reflected in the structure of the food-gathering mechanisms and alimentary canal. pH changes along the course of the gut are small and the fluid there is usually around the neutral point. Proteins are usually digested extracellularly by proteinases, which are strong in the carnivores and are seemingly of a tryptic nature. A selection of carbohydrases is generally present, and these are particularly comprehensive in the echinoids where a variety of polysaccharides may be dealt with, possibly with the assistance of bacteria, although their role is not clear. Lipases are generally intracellular, but the exact site of phagocytosis is variable, being either amoebocytes or the endothelial lining of the gut. Traditionally the major organs for the elaboration and transport of enzymes and the products of digestion in the echinoderm body, the amoebocytes are now being investigated with modern techniques and as a result their role is open to question. Food storage, chiefly in the form of glycogen and lipids, is in the caecal walls and gonads. In animals lacking caeca, the stomach and body wall serve as storage organs.

The bionomic importance of the group can be seen in the role of the holothurians in the turnover of organic material in the upper layers of the littoral substratum or in the grazing bare of large areas of rock surface or corals by echinoids and asteroids. In competing for food with animals which are directly consumable by man, asteroids play an important role in addition to their direct consumption of shellfish. Only rarely are echinoderms themselves used for food in such ways as trepang.

CHAPTER 2

EXCRETION AND THE ROLE OF AMOEBOCYTES

FROM what has been said about the processes of feeding and digestion, it will be apparent that echinoderms have a very simple metabolic economy and operate at a low level of activity. The quantity of nitrogen to be excreted would, therefore, be expected to be quite small. Being isotonic with the surrounding sea water, they are not under any osmotic stress and the external supply of water is in no way limited. It is, therefore, possible for nitrogen to be excreted as ammonia without its concentration attaining a toxic level within the body. The whole phylum lacks any morphologically differentiated excretory organs and there are thus two possible mechanisms by which excretion can take place. Firstly, simple diffusion will permit excretory products like ammonia to pass into the external medium via the general body surface; secondly, more refractory substances can be eliminated in solid form. In both these methods amoebocytes may play a critical role.

Crinoidea

In the Crinoidea nothing is known about the first method of excretion or the exact form which the nitrogen takes, but amoebocytes are commonly found in all the tissues. Cuenot[179] suggests that these transport waste materials into the connective tissue filled spaces within the body, which eventually become loaded with brown granules and impart a darker colour to the older animals.

Holothuroidea

As has already been noted, the holothurians have a large complement of amoebocytes and it is usually stated that such cells, loaded with the excretory material which they have ingested, migrate through the walls of the respiratory trees and pass out with the exhalent current.[42,53,513,706] In the apodous forms such as *Leptosynapta,* which lack respiratory trees, the amoebocytes pass into reniform structures called urns and thence into the body wall.[346] Soluble dyes injected into the coelom make their exit by this route and also into the lumina of the gut and gonad. No analytical data are available to indicate the nature of the presumed excretory granules, nor is there any estimate of the quantities involved. However, there is no doubt that a certain amount of soluble nitrogenous waste is produced by these animals. Such wastes have their origin in two main sources: from protein and amino acid metabolism and from the purines and pyrimidines of nuclear metabolism. Nitrogen from the first of these sources may be excreted directly as amino acids or ammonia resulting from the various forms of deamination. These two substances form the bulk of the non-protein nitrogen (NPN) found in water in which *Holothuria tubulosa* had been kept for 24 hr.[194] Primitively, nitrogen from purine metabolism is also converted into ammonia by a somewhat lengthy enzyme system. Biochemical evolution in this sphere is characterized by a loss of one or more of these enzymes with the result that a variety of intermediate compounds may be excreted. Conflicting results have been published regarding the nature of these compounds. The claim that some urea is excreted by *Holothuria,*[194,688] has not been confirmed in *Thyone*[334,782] The absence of uric acid, creatine and creatinine was established, the importance of which will be referred to again.

Echinoidea

In the Echinoidea the coelomocytes also appear to play an important role in the excretion of insoluble or particulate material. Carmine, carbon particles and other inert matter injected into the coelom are quickly ingested by these cells which then migrate through the thin-walled regions of the body, such as the gills. Alternatively, they may deposit it in the body wall or axial organ. The coelomic epithelium itself may also phagocytose particles, which is interesting in view of this epithelium being the origin of certain types of coelomocyte.[26,73,402,421]

Experiments of this nature must be viewed with some caution, for many of these substances may not be as inert as is often thought and even their mechanical presence could elicit phagocytotic activity in normally quiescent cells. Under normal conditions coelomocytes, loaded with yellow or brown inclusions and which are generally regarded as excretory in nature, appear on the outside of the gills.[212b] The soluble nitrogenous material in the coelomic fluid appears to consist of ammonia, amino acids, and urea with little or no uric acid.[150,194,560,688] In *Strongylocentrotus francescanus* the presence of creatine and creatinine has also been detected in the coelomic fluids but amino acids are not reported.[560] In the closely related *S. purpuratus* a comprehensive list of a dozen amino acids found in the coelomic fluid has been published.[302] Uric acid was detected in the coelomic fluid and intestine of *Arbacia punctulata* and it was concluded that excretion occurred by way of this organ.[782] Florkin and Duchateau[252] point out that as sea urchins possess uricase, the soluble excretory products will be principally urea and ammonia and this seems well authenticated by many other workers. For example, some 60% of the total NPN excreted by *Diadema antillarum* is ammonia and another 30% is amino acid.[467] No urea, uric acid or other purine bases were detected, but the ammonia content increased along the intestine and rectum and such evidence supports the earlier suggestion that the hind gut has an excretory function.

Asteroidea

As early as 1888 Durham[212a] using *Asterias rubens* indicated that the ingestion of foreign particles by coelomocytes occurred. This is followed by their migration through the dermal branchiae, the tips of which have even been observed to constrict, carrying off many cells with them.[38,177,471] These cells have also been observed, loaded with refringent granules, possibly purine in nature, making their exit across the walls of the caeca and madreporite[133] and across the bursal walls in ophiuroids.[176] In those ophiuroids lacking bursae these granules are deposited in the tissues and body wall. The injection of soluble dyes has yielded little information beyond the fact that the various cells and tissues have a differing affinity for such dyes. With the constant implication of the coelomocytes in the excretory mechanism, it has often been proposed that the organs responsible for the production of these cells are themselves concerned with excretion. Tiedemann's bodies have been supposed to be one source of coelomocytes and the axial organ another, although evidence for this is uncertain.[177] Transplantation of supernumerary axial organs resulted in a slight enlargement of the indigenous organ and extirpation in a slight enlargement of the Tiedemann's bodies.[788] A direct excretory function for these structures seems unlikely and their function will be reconsidered below.

Ammonia and amino acids are the chief soluble nitrogenous wastes of *A. rubens* being present in both the coelomic fluid and the water in which the animal has been kept for 24 hr.[194] An excess of ammonia in the external medium suggests some kind of selective secretory mechanism may exist within the integument. The excretion of considerable quantities of amino acids by the sea urchins *Paracentrotus* and *Strongylocentrotus* was also noted. Cohnheim[150] found a rise, only in the total NPN concentration of the medium, whilst others have found most urea and ammonia, for example, in the coelomic fluid of *Picnopodia helianthoides*. Creatine and creatinine have been detected in *Pisaster ochraecous* whilst a detailed analysis of the amino acid content of the closely related *P. brevispinus* has been made.[302] No ammonia, urea or creatine were, however, to be found in *A. forbesi*,[782] but uric acid first demonstrated in the pyloric caeca by Griffiths in 1888[314] was confirmed. The enzymes uricase and allantoinase, both important in nitrogen metabolism, have been found in *A. rubens*,[261] so it seems likely that the quantities of purine bases will be quite small as they will be excreted as urea, ammonia and even allantoic acid. Such may well form part of the "undetermined" fraction so often mentioned in analytical data.

Ophiuroidea

The nature of the soluble excretory products in the Ophiuroidea is unknown.

There is thus a good deal of conflicting information in the field of nitrogen metabolism of echinoderms. Whilst it is to be expected that ammonia would be the most common form of nitrogenous waste, instances have been cited where even the presence of this ubiquitous substance is a matter of controversy. Amino acids also contribute, sometimes fairly substantially, to the excretory NPN fraction. It is obviously of great importance in studies of this kind to know several things about the condition of the experimental animals, such as the nutritional state or stage of the reproductive cycle. In most of the published work such data are frequently omitted. The evolutionary loss of enzymes in the system responsible for purine metabolism does not appear to have proceeded very far in echinoderms. Uricase and allantoinase have been demonstrated in the adult sea urchin and starfish, but in the embryo the complete sequence is present.[92] A reinvestigation of the whole subject of nitrogen metabolism, with particular emphasis upon the enzyme systems involved, would seem desirable. The role of the amoebocytes, labelled with radioactive nitrogen compounds should also prove a revealing study.

Origin of Coelomocytes

Consequent upon the acknowledged transport functions of the perivisceral fluid, authors have frequently been led to refer to it, incorrectly, as blood. The main body cavity, together with those of the water vascular and haemal systems are, however, coelomic in nature, although like true blood, the perivisceral fluid consists of a liquid phase, the plasma and a cellular component. The origin of this cellular component is diverse and several sites of formation have been suggested, including the peritoneal wall, walls of other coelomic cavities and mesenchyme cells, as well as the more obvious lymphoid tissue such as Tiedemann's bodies. Schinke[701] replaced the coelomic fluid of *Psammechinus miliaris* with sea water and found that 60% of the coelomic cells were reformed within 24 hr and the process was complete in from 2 to 4 days. In a more recent study it was concluded that some, if not all, of the lymphocytes originate from mesenchymal cells in the haemal vessels and that these possibly differentiate into other forms.[346] Some species possess a range of coelomocytes which can be arranged in a sequence suggestive of such a process of differentiation.[599] Experiments involving the injection of tritiated thymidine showed that all coelomocytes except the vibratile cells in *S. purpuratus* became radioactive.[354] As other sites of suggested coelomocyte formation were also labelled, these experiments probably indicate regions of high thymidine turnover rather than providing evidence for the genesis of cell types. Cells similar to those found in the coelomic cavities are to be observed in most of the tissues of the body. They are obviously free to wander throughout the body but the conditions governing their movements are unknown. Chemical gradients such as might arise from variations in metabolic activity could provide an explanation.

Coelomocyte Types

A detailed description of these cells would be out of place here, but a brief mention of the variety of forms which exist will be helpful when considering one further important function which they perform, namely, clotting. In the Crinoidea, a class noted for the abundance of coelomocytes, there appears to be two basic types, although naturally there is a good deal of variation between the species. An amoeboid form, which is probably phagocytic, and a granular form, which is most likely to be concerned with the transport of food materials, are amongst those figured by various authors.[175,176,317,658] A wide range of forms are to be found in the Holothuroidea and the situation is complicated by the varying nomenclature given to them by different authors. The cells include haemocytes, which are red in colour and contain a form of haemoglobin,[357] various phagocytes, and amoebocytes with both coloured and colourless granules. Some possess crystalline inclusions which have been associated with excretory processes, whilst others possess only vacuoles or vesicles. Like holothurians, echinoids possess a somewhat similar variety of coelomocytes. Pigment cells are often present and commonly contain echinochrome, whilst other colourless cells may contain inclusions associated with pigment formation or transport (see Chapter 7). Such inclusions have been shown, for example, to be the precursors of melanin, the pigment itself being formed by the breakdown of these cells in the body wall.[529] Apart from a variety of non-phagocytic forms, there are those with either slender or broad pseudopodia which are strongly phagocytic. It seems likely that these two forms constitute different phases of the same cell.[598,599] Intense phagocytosis of the macrophages has been observed in *Caudina chilensis* when this animal was kept for any length of time in an aquarium. Vibratile coelomocytes are also common and are thought to be detached coelomic epithelial cells which have become amoeboid,[469] and are thought to play a part in the transport of nutrients especially in those animals where the nutritive content of the body fluid is low.

Much less variation is encountered in the coelomocytes of the Asteroidea, there again being only two basic types. Actively phagocytic amoebocytes with either long slender pseudopodia or broader ones, as before, constitute two phases of the same cell and are the ones responsible for the ingestion of foreign particles.[212a] The remainder consist of small spherical corpuscles. Granular cells with numerous slender pseudopodia are to be found in the coelomic cavities of the Ophiuroidea. Modern phase contrast microscopy has done much to simplify the diversity of descriptions of these cells, especially regarding the transformation of one cell type into another. (For reviews, see refs. [3, 18,76,345,701].)

Clotting

Clotting in echinoderms was first investigated as long ago as 1880 by Geddes,[287] but further work was sporadic until the more recent systematic investigations,[4,82,105] asteroids and echinoids having received the more thorough attention. Although all the cells mentioned above are involved in the formation of the clot which forms at the site of an injury, only two types are responsible for the initiation of the clot. These are the amoebocytes with long slender pseudopodia, termed filiform amoebocytes, and those cells termed hyaline haemocytes which occur only in the Echinoidea. Burton,[105] however, distinguishes three morphological types—bladder amoebocytes which are phagocytic and are involved in coagulation; spherule amoebocytes which may be either pigmented or unpigmented; and flagellated cells. Three fairly distinct types of clot have come to be recognized. The simplest type consists of a cellular aggregation in which the integrity of the participating cells is maintained. Later the cells disperse, such clots being found in the crinoid *Heliometra*, the ophiuroid *Gorgonocephalus* and the clypeastrid *Dendraster*. The formation of such clots has been shown to be dependent upon the presence of calciumions and a tissue factor released by the damaged tissue.[73,82,189,202] Additionally it has been shown that the damaged tissues of animals from certain other phyla could provide the necessary factor. Clotting incidentally is often associated with pigment liberation and granule breakdown. From this stage it is but a short evolutionary step to the formation of the true plasmodial clot which occurs in a number of species.[178,287,358,402,403,752] Significantly it has been shown that the clots formed by asteroids are independent of the presence of calcium ions, but that their formation is inhibited by the presence of reducing agents or heavy metal ions which specifically bind sulphydryl groups.[82] The addition of an oxidizing agent to the former system or chelation of the metal ions in the latter case, brings about a reversal of the inhibition. It is clear that the conversion of sulphydryl groups to disulphide linkages plays an important part in clot formation. Echinoids generally form a third type of clot, namely a fibrous one, also involving cellular agglutination, which has been described in a number of species.[82,176,178,314,402,403,513,698,752,753] The physiological results were similar to those obtained upon asteroids, but with the additional possibility of the participation of the plasma proteins in the strengthening of the clot. A fall of between 15 and 25% in the total plasma nitrogen during the process of clotting was demonstrated. Starvation, which also results in a fall in the plasma nitrogen, impairs the clotting reaction. Coelomic cells migrate rapidly to the site of injury[427] and many, although not participating in wound healing, such as the red-brown amoebocytes, quickly gather at the site of injury. The mass which forms the cellular basis of the clot undergoes contraction and dilation while cells carrying calcium carbonate migrate from the deeper regions of the animal.

There is a possible ecological significance in the distribution of these types of clotting mechanism. An animal like an asteroid or holothurian can contract its body wall at the site of injury to prevent or reduce the loss of coelomic fluid. These two groups have the poorer clotting and the chemically more simple perivisceral fluid, especially with regard to the amino acid composition. In the echinoids with their hard test no such contractions of the body wall are possible and these animals possess a much more efficient clotting mechanism and also the chemically more sophisticated perivisceral fluid.[302,701]

Bacteriologically, the perivisceral fluid is normally a sterile medium and it is likely that the phagocytic amoebocytes are responsible for maintaining this condition.[38] Bacterial infection of the perivisceral fluid has been shown to be followed by numerous symptoms, including muscular weakness, sluggishness and agglutination of the amoebocytes. Bacteria have been found to be present after autotomy and in animals living in stagnant conditions, which may well be one of the reasons why echinoderms do not usually live for any length of time in small aquaria.

The ubiquity of the amoebocytes is one of the characteristic features of the echinoderms. Although suspended in a coelomic fluid rather than a blood of more nutritive nature, they perform many of the functions of both red and white corpuscles of other animals. Ingestion of foreign particles, including bacteria, coagulation to form a clot after injury to the animal coupled with the thrombocyte-like behaviour of the hyaline haemocytes, the occasional possession of haemoglobin and the carriage of waste materials, food substances and possibly enzymes are all points of comparison and for further study. Such a list of parallel functions is, however, surely impressive.

Fig. 4. Agglutination in the asteroid *Pisaster ochraceous*. × 47. A true plasmodium is seen since no cell membrane is found between the agglutinated cells. Several cell nuclei can be seen (arrows).

(*facing page 32*)

CHAPTER 3

SALINITY TOLERANCE AND OSMOREGULATION

WHEN considering the nitrogen metabolism of echinoderms, it was pointed out that members of this phylum lack any morphologically differentiated excretory organ. Such a condition would rule out the possibility of any significant regulation of the osmotic concentration of its coelomic fluids, which may constitute 30% or more of the total

TABLE 1. EURYHALINE ECHINODERMS

Species	Location and salinity	Reference
ASTEROIDEA		
**Solaster* sp.	Cape Shelagsk, U.S.S.R., $24\cdot5\,^{0}/_{00}$	198
**Asterias rubens*	Baltic Sea, $8\cdot0\,^{0}/_{00}$	91
A. forbesi	Long Island Sound, U.S.A., $18\cdot0\,^{0}/_{00}$	477
**Marthasterias glacialis*	Black Sea, $18\cdot0\,^{0}/_{00}$	762
OPHIUROIDEA		
**Ophiothrix fragilis*	Black Sea, $18\cdot0\,^{0}/_{00}$	839
**Ophiura albida*	Baltic Sea, $8\cdot{}^{0}/_{00}$	708
Ophiophragmus filograneous	Cedar Key, Florida, $7\cdot7\,^{0}/_{00}$	745
Ophiocten sp.	Cape Shelagsk, $24\cdot5\,^{0}/_{00}$	198
ECHINOIDEA		
**Echinus acutus*	Black Sea, $18\cdot0\,^{0}/_{00}$	762
HOLOTHUROIDEA		
Stereoderma kirschbergi (*C. orientalis?*)	Black Sea, $18\cdot0\,^{0}/_{00}$	134
**Leptosynapta inhaerens*	Black Sea, $18\cdot0\,^{0}/_{00}$	27

*Known only from full salinity sea water for the remainder of their European distribution.

body weight. Consequently, it is to be expected that echinoderms will be an almost exclusively marine phylum. This generalization is true for the vast majority of the group, but in the last twenty years a significant number of examples of echinoderms entering water of reduced salinity has been recorded in the literature (see Table 1).

Salinity Tolerance in the Field

Although not recorded from brackish water, the crinoid *Tropiometra carinata* seems indifferent to salinity changes between 75 and 110% sea water, but some pigment was lost from the integument at 120% sea water.[141] The ability to tolerate brackish water is most marked in asteroids and ophiuroids. Euryhaline asteroids are found in widely separated geographical regions; for example, *Asterias rubens* penetrates the Baltic Sea as far as the eastern end of Rugen Island, where the salinity is only 8‰.[91,708] The closely related *A. forbesi* penetrates Long Island Sound as long as the salinity does not fall below 18‰.[477] The most euryhaline echinoderm so far recorded is *Ophiophragmus filograneous*, which lives off the coast of Florida in water of salinity of 7·7‰.[754] Many other ophiuroid species have been recorded from the Black Sea, together with holothurians and an echinoid, where the bottom salinity falls to around 18‰. The hard test of echinoids could reasonably be expected to preclude them from diluted sea water, but other than this, it is difficult to discern any common features in this assemblage of animals. Geographically, it would seem that they occur only in those regions where the dilution of the sea has been gradual, taking a number of geological epochs, rather than in those regions subject to seasonal or even diurnal fluctuations. The former regions might provide the appropriate conditions for the slow process of acclimatization to take place, rather than in the more rapidly varying conditions of an estuary. Both the Baltic and the Black Seas would fulfil these conditions and it is interesting to note that most of the species listed as occurring in these Seas do not enter water of reduced salinity elsewhere in their range. For example, *A. rubens* does not enter water of significantly reduced salinity around British coasts[60] and neither does *Ophiura albida*. This sporadic penetration of diluted sea water within a single species suggests the possibility of the existence of physiological races. Such races would require considerable periods of time to evolve, a factor which would also point to their origination in larger bodies of water. Recently, two such races of the echinoid *Psammechinus miliaris* have been found to exist off the Swedish coast:[292] a Z-form, which inhabits the littoral zone and is tolerant of widely divergent salinities from 16 to 32‰, and an S-form, with a narrow tolerance of between 30 and 34‰ and which is restricted to deeper waters. Attempts to adapt either form to the salinity range of the other have so far failed.[470] The cleavage rate and salinity optima for the eggs followed that of the parents and not that at which the gonads had been matured. This example shows how the evolution of differing salinity tolerances could be a factor in speciation, leading perhaps, as in the case of *P. miliaris*, to geographical isolation.

Under field conditions many echinoderms must, temporarily at least, experience an environment which is significantly less saline than normal without permanent injury. For example, many species must be exposed to the effects of rain when stranded by the tide, for instance, *Strongylocentrotus purpuratus*.[296] Fresh water run-off, particularly in rainy seasons or periods of melting snow or ice, can dilute the environment of littoral species. Animals shown to be tolerant of such conditions include the holothurian *Opheodesoma spectabilis*[270] and the Antarctic asteroid *Odontaster validus*.[619] Sudden inundations such as that recorded at Newport Bay Harbour must also severely dilute the environment; nevertheless, both the ophiuroid *Amphiodia barbarae* and the echinoid *Dendraster excentricus* seemed to survive.[564] The latter species may have avoided the full force of the fresh water on this occasion by burrowing; however, the salinity of interstitial water, particularly if this is influenced by seepage, will be of great importance to such animals.[726]

Taken all together, this body of evidence suggests that echinoderms may not be quite the rigidly stenohaline group they are traditionally considered to be. They are to be found over a sixfold range of salt concentration, from 8‰ in the Baltic Sea to 46‰ in the Red Sea. Great importance is attached by palaeontologists to the appearance of echinoderms in the fossil record as being indicative of a fully marine environment. Perhaps a little more critical approach may not be out of place in this context, although it should not be inferred that any ancestral echinoderm was an inhabitant of fresh water. D'Iakonov[198] took a substantially similar point of view, after studying a more limited range of animals.

Experimental Tolerance

Species not normally occurring in diluted sea water have often been made the subject of experiments to determine their lower limit of salinity tolerance and whether this value differs markedly from those animals which normally enter brackish water. In this respect, asteroids and echinoids have been most commonly investigated. The lower limit of salinity tolerance of *Asterias vulgaris* was found to be 14‰,[716] a result which was later criticized on the grounds that the acclimatization period was of too short a duration for a genuine limit to have been obtained.[477] The limit for the closely related *A. forbesi* was found to be higher, namely, 18‰, a value later confirmed,[806] when it was also shown that within the experimental period this limit was unaffected by the rate of dilution of the medium. *A. rubens*, the common littoral equivalent on this side of the Atlantic Ocean, could not however tolerate sea water more dilute than 23‰.[60] The echinoids *Arbacia punctulata* and *Lytechinus variegatus* were found to have tolerances of 19‰ and 23‰ respectively,[806] whilst the value for the related *S. droebachiensis* was 21·5‰.[445]

To maintain a viable population at such salinities, it is obviously necessary for the animals to be able to mature, reproduce and for the larvae to develop in such conditions. None of the artificial acclimatization experiments has covered the complete life cycle. Sagara and Ino[685] found that the larvae of the Japanese star fish *Asterias amurensis* could tolerate salinites of 14·8‰ to 42‰, but no data upon the adult's ability to mature and spawn at these salinities were given. The rapid death of the echinoderm larvae in the plankton when they encounter water of significantly reduced salinity indicates the great susceptibility of this developmental stage.[239,757] Salinity is thus an important factor controlling the distribution of a number of echinoderms. From ecological surveys it was concluded that *A. vulgaris* and *S. droebachiensis* are prevented from entering estuaries solely as a result of their inability to tolerate greatly diluted sea water.[761] Pearse in 1908[616] came to the same conclusion about *Thyone briareus*. Chemical differences present in water in insufficient amounts to affect the osmotic pressure may, however, be quite sufficient to influence an animal's distribution. Echinoderms generally seem to be more characteristic of water containing the arrow worm *Sagitta elegans* than the related *S. setosa*.[781] It is conceivable, therefore, that small changes in water currents could well influence local echinoderm populations. Vinogradova[798] even thinks that elements such as copper, manganese, iron, etc., could also be important in this respect and could even influence their migration.

Effects of Reduced Salinity

Reduced salinity can have a number of physiological as well as morphological effects upon echinoderms. In those species living in permanently brackish water, Thorsen[757] has noted a reduced "spawning intensity", for example, in *A. rubens* which at the extremes of its range may not even breed at all. The population in such regions is most likely maintained by a migration of adults or a drift of larvae, although, as already noted, these larvae are amongst the most susceptible to lowered salinity. The diameter and number of ova produced by this species at Kiel (16‰) is also adversely affected, as is the latent period and reaction time for the characteristic righting reaction which these animals perform when placed upon their aboral surface.[65,422] Differences in the gross chemical composition between Kiel and North Sea (30‰) animals have also been recorded (see Chapter 4), but generally speaking the animals from the Baltic Sea have a higher water content, but a lower ash, carbohydrate and lipid content.

Several of these effects can be reproduced in animals immersed in diluted sea water, whether or not they will naturally enter water of reduced salinity. Swelling, increase in weight and loss of pigment from the integument, which often becomes blotchy due to localized destruction of the epidermis, are the most obvious symptoms. Possibly as a mechanical effect of their turgidity following immersion in diluted sea water, the animals are immobilized or alternatively the reason may be a more fundamental physiological one. After a few days in 60% sea water, *A. rubens* from Roscoff lost some of its turgidity and regained its normal level of activity.[373,374] This author has kept North Sea representatives of this same species for a similar length of time in 70% sea water without observing any significant improvement in the animal's activity or ability to right itself and no decrease in weight whatsoever. After a week in diluted sea water *Strongylocentrotus droebachiensis* also seems to lose some of its turgescence and regains its activity.[445] Prolonged immersion in too great a dilution is fatal, but some animals can withstand very dilute sea water for short periods. *A. forbesi*, for example, could withstand up to 2 hr immersion in water of only 3‰ without any significant mortality after return to normal sea water.[477]

Water Relations

The increase in weight due to the osmotic influx of water when animals are placed in diluted sea water has been recorded many times.[285,290,340,341,343,649,702] Bethe examined the time course of the weight changes and showed that the initial weight of *A. rubens* and *Echinus esculentus* rose by some 10% in 75% sea water and this was followed by a return to the original weight in from 6 to 24 hr.[56] Similar results also with *A. rubens* were obtained by Maloeuf.[483] Both authors found that when weight equilibration was reached, the perivisceral fluid was isotonic with the medium. Thus they showed that some species of echinoderm were at least capable of weight regulation or control of their coelomic volume in diluted media. However, owing to the greatly increased turgescence and fragility of the integument under these conditions, the body wall is liable to rupture when removed from the medium for weighing. By taking special precautions during the weighing procedure of such distended starfish the weight was found to remain constant for up to 48 hr without any rupture of the integument and consequent loss of coelomic fluid. It is concluded that this species was not capable of any form of weight regulation, a conclusion unaffected by size, sex, state of the breeding cycle or environmental temperature.[60]

Weight changes in sea water more dilute than 50% are somewhat erratic, *A. vulgaris* swelled enormously in tap water,[285] but *A. rubens* took up but little water in this medium, doubtless due to the rapid loss of salt.[60] Conductivity measurements indicate that by far the greatest salt loss occurred from holothurians when these animals were put in distilled water, a pathological experiment rather than a physiological one.[384] Recently two instances of echinoderms being able to some extent to control their body volumes have appeared in the literature. As already noted, the holothurian *Opheodesoma spectabilis* is tolerant of a certain amount of fresh water run-off,[270] and interestingly this animal has a thick mucus covering which could form a protective layer as it does in the eel. When placed in diluted sea water the animal did not gain weight as expected and in some instances even lost weight. If the animal was drained of its coelomic fluid and replaced in sea water it increased in weight, almost up to its original level. Closure of the mouth or anus indicated that water re-entered through the mouth. The wet weight of individual specimens of the Antarctic asteroid *Odontaster validus* were observed by Pearse[619] to vary by up to $\pm 30\%$. These changes often exhibited a short-term periodicity of between 10 and 20 days, and seasonal variations were also recorded which do not appear to be correlated with any variations in the body component indices. When placed in dilute sea water there was little change in weight, but all of them lost weight when returned to water of normal salinity. This would suggest that they had achieved isotonicity with the diluted medium by loss of salt, although no measurements of osmotic pressure were taken at this point. This starfish is of the cushion-star variety, many of which have a well-known tendency to inflate and deflate.[674] This animal can also withstand the effects of fresh water run-off in its shallow water environment and its ability to control its coelomic volume may be of value in this respect. Pearse also noted that when narcotized with magnesium chloride, freshly caught turgid specimens lost weight, regaining their original weight upon return to normal sea water. When *A. rubens* was placed in mixtures of sea water and isotonic glucose or choline chloride this animal too lost weight. From analysis of the body fluids during the course of this experiment, it was clear that both water and ions in the same proportion as in normal coelomic fluid has passed out across the body wall.[61] When returned to normal sea water these animals gained weight, although their coelomic fluids had not been made hypertonic by the influx of glucose, and approximated to their original weight. There are thus several examples of animals deprived of their coelomic fluid by various means all returning to something approaching their initial weight in normal sea water. It would appear that such animals might have an optimum coelomic volume to body volume ratio to which there is a tendency to return when this ratio is seriously disturbed, a kind of "physiological weight card". Our knowledge of the weight regulatory abilities, especially in diluted media, of this whole phylum are restricted to one or two species. Further investigations, particularly with the more euryhaline ophiuroids, would be invaluable.

Hypertonic Environments

Adaptation to a hypertonic medium has received but little attention. When placed in hypertonic sea water the coelomic fluids of *Marthasterias glacialis* and *Holothuria tubulosa* attained the same depression of the freezing point and conductivity as the medium, similar results being obtained by Koizumi working with *Caudina chilensis*.[411] Dakin placed *E. esculentus* in hypertonic sea water, depression of the freezing point −2·98°C, and found that equilibrium was nearly obtained after only 3 hr, the depression of the freezing point being −2·07°C.[184] A similar experiment with *A. rubens* resulted in an almost complete loss of coelomic fluid after this period of time.

The large changes in weight, such as those mentioned above, involve the passage of considerable quantities of water across the body wall. Work at the beginning of this century, especially that by Henri and Lalou,[340–343] indicated that not only was the body wall of echinoderms extremely permeable to water, but that it was in fact truly semi-permeable. This view was later modified when analysis of the stomach fluid of holothurians showed that it contained less chloride than did the coelomic fluid. Various poisons, such as chloroform and sodium fluoride, were shown to produce relatively large changes in the permeability of these animals. Modern ideas concerning the active nature of the transport of various substances across cellular membranes has led to a more dynamic approach to the problems of the osmotic adjustment and permeability of echinoderms.

No direct measurements are available for an absolute value for the permeability of the body wall of echinoderms to be ascertained. Once again, experiments upon *A. rubens* suggest that both the oral and aboral surfaces, each bearing its own form of thin-walled projection, contribute about equally to the passage of water[60] and oxygen.[509] Furthermore, it has also been demonstrated that water can pass easily

TABLE 2

OSMOTIC PRESSURE OF ECHINODERM BODY FLUIDS UNDER ENVIRONMENTAL CONDITIONS

Species	Fluid	Δ(°C)	mM/l Cl⁻	Reference
ASTEROIDEA				
Astropecten aurantiacus	Amb.	−2·312	—	86, 88
Patiria miniata	PV	−1·885	—	
	SW	−2·075	—	367
Asterias rubens (Kiel)	PV	—	255	
	SW	—	252·8	707
A. rubens (North Sea)	PV	−1·86	—	
	Amb.	−1·92	—	
	SW	−1·90	—	60
Marthasterias glacialis	PV	−2·65	—	
	SW	−2·68	—	269
ECHINOIDEA				
Echinus esculentus	PV	−1·86	—	
	SW	−1·86	—	184
Strongylocentrotus droebachiensis	PV	−1·776	—	
	SW	−1·759	—	152
HOLOTHUROIDEA				
Holothuria sp.	PV	−2·312	—	
	SW	−2·195 to −2·360	—	87
Cucumaria frondosa	PV	−1·750	—	
	Amb.	−1·749	—	
	SW	−1·759	—	614
Caudina chilensis	PV	—	528	
	SW	—	529	411

Amb., ambulacral fluid from the water vascular system.
PV, perivisceral fluid.
SW, sea water.

from one coelomic cavity to another. The stomach wall is also extremely permeable to water and in the process of equilibration to a medium of differing osmotic pressure it is possible that water could enter or leave the animal by this route. Starfish stranded by the falling tide frequently have their stomachs everted, penetrating some little distance down between the gravel. At the lower levels of the beach, the water table is usually only a few centimetres below the surface and it would seem possible that water lost by evaporation from the aboral surface could be replaced osmotically from this source. Animals stranded higher up the beach on dry gravel quickly become desiccated.

The picture gained thus far of the water relations of echinoderms is of an animal enclosed by a body wall which is generally very permeable to water, and made up of a number of coelomic compartments between which water can move freely. The actual osmotic pressure of the fluids within either the perivisceral coelom or the water vascular system has been measured by a large number of workers over the years using a variety of techniques. Table 2 gives a selection of their results. Again asteriods have been one of the most intensively worked groups, especially *A. rubens*. As mentioned above, this species ranges from full salinity sea water to brackish water of 8‰. Although no osmotic pressure measurements seem to have been made upon this animal at the extremity of its range, those made by Seck[707] upon animals from Kiel (16‰) show that it is still isotonic with the medium. It is a regrettable omission that no data are available from the Crinoidea or Ophiuroidea, where, unfortunately, the volume of coelomic fluid is extremely small.

Intracellular Osmotic Pressure

Although there seems to be no control of the osmotic pressure of the body fluids in echinoderms, there would appear to be some control of the intracellular osmotic pressure and much recent work has centred around this aspect. Invertebrates generally have a high intracellular concentration of amino acids and as long ago as 1959 it was shown that such was true for the starfish *Luidia* and the holothurian *Thyone*, glycine and taurine being the main components.[714] They suggested that these substances served an osmoregulatory function. The levels of these substances in different tissues of the echinoid *S. purpuratus* have also been examined.[405] Several workers have recently recorded a restoration in activity coupled with a reduction in the water content of the tissues after prolonged immersion in diluted sea water. Using *A. rubens* small variations between the wet weight of the gut caeca in full salinity sea water and in 60% sea water was thought to represent some sort of intracellular osmotic pressure control mechanism.[250,251,373,374] Variation in the intracellular concentrations of free glycine and free taurine play an important part in this mechanism, their decrease sparing the other cellular components from larger fluctuations. The decrease in the intestinal concentration of ninhydrin positive substances in *S. droebachiensis* was an almost linear relation to the salinity. Due to this osmotic buffering the intracellular salinity is maintained within closer limits than the medium.[445] However, the concentration of taurine remained constant and low whereas in the mussel *Mytilus edulis* it rose, exerting a sparing action upon the amino acids.[444] The variation in the concentration of amino acids within the cell is possibly due to a shift in the equilibrium between the depolymerized and the polymerized forms, possibly reflecting a balance between metabolism and excretion, as was suggested by Potts.[637]

An interesting observation has been made upon the radial nerve cords of the starfish *A. glacialis*, which suggests that some over-riding central control of such reactions may be within the bounds of possibility.[629] It was observed that the number of granules in the nerve cords varied with the external salinity, disappearing completely under hypotonic conditions. Unger[779,780] had already suggested an osmoregulatory function for granules which he observed in a similar situation. Florkin has made the distinction between animals which have evolved a mechanism for the control of the osmotic pressure of their extracellular fluids, namely, excretory organs, and those which have evolved only an intracellular mechanism, such as is seen here in the echinoderms.

CHAPTER 4

IONIC REGULATION AND CHEMICAL COMPOSITION

THE accurate chemical analysis of body fluids is one of the many aspects of biological research which has followed upon the innovation or development of techniques in other sciences. Although a considerable body of data upon echinoderms now exists, it has been accumulated in a more or less random fashion over a period extending back into the last century. Consequently, such information is often of varying accuracy and once again data from crinoids and ophiuroids are regrettably lacking, the more so in the latter case as these animals are amongst the most euryhaline of echinoderms. The composition of fluid within the other coelomic cavities, such as the water vascular system and haemal system, have been comparatively neglected. A summary of the more complete analyses can be found in Table 3.

Ionic Regulation

Briefly, it will be seen that potassium is accumulated in the perivisceral fluid of asteroids from 9 to 16% above that of the environment. This would appear to be maintained both in normally euryhaline animals[707] and in those which have been experimentally acclimatized to reduced salinity.[61] Calcium, too, is also accumulated to a slight extent and conversely there is a small reduction in the magnesium concentration. In echinoids, however, there seems to be little or no difference between the internal and external potassium levels, although a small and possibly insignificant sex difference has been observed, the female containing the slightly greater amount.[636] There is, however, a rise in the calcium concentration, in one instance by as much as 57%[560]

TABLE 3. IONIC COMPOSITION OF ECHINODERM BODY FLUIDS

Species	Fluid	Na	K	Ca	Mg	Cl	SO_4	pH	CO_2	Reference
ASTEROIDEA										
Astropecten sp.	PV	—	14·3	14·2	53·9	659	32·5	—	—	58
	SW	—	12·3	12·7	61·7	627	31·2	—	—	
Solaster endica	PV	420	9·7	9·6	49·8	488	30·0	6·9	2·40	614, 152
Asterias vulgaris	PV	460	8·3	9·0	30·7	505	25·4	7·54	2·13	
	SW	458	8·5	9·7	33·8	492	25·4	8·1	2·15	
A. rubens (Kiel)	PV	216	5·44	5·60	24·22	255	13·05	—	—	707
	SW	215	4·98	5·56	24·05	252·8	13·08	—	—	
A. rubens (North Sea)	PV	428	9·5	11·7	49·2	487	26·7	7·2	2·82	61
	Amb.	418	15·1	9·7	50·3	481	25·5	6·9	—	
	SW	429	9·5	10·8	49·0	494	25·4	7·8	2·52	
A. rubens (North Sea animals acclimatized to similar salinity as Kiel)	PV	234	5·9	6·0	27·1	265	13·7	—	—	
	Amb.	234	11·8	4·8	28·1	264	13·0	—	—	
	SW	232	5·2	5·0	27·0	272	14·0	—	—	
Marthasterias glacialis	PV	459	10·9	10·2	51·2	540	27·6	—	—	665
	Amb.	—	11·9	—	—	—	—	—	—	
	SW	459	9·8	10·1	52·4	536	26·7	—	—	
ECHINOIDEA										
Echinus esculentus	PV	530	13·4	12·8	50·6	546	—	—	—	57
	Amb.	521	19·3	15·1	49·8	552	—	—	—	
	SW	487	13·6	10·5	—	525	—	—	—	

Abbreviations: PV, perivisceral fluid; Amb., ambulacral fluid from the water vascular system; SW, sea water.

TABLE 3. IONIC COMPOSITION OF ECHINODERM BODY FLUIDS—*cont.*

Species	Fluid	Na	K	Ca	Mg	Cl	SO_4	pH	CO_2	Reference
Paracentrotus lividus	PV	—	12·7	12·7	51·8	621	32·1	—	—	58
Sphaerechinus granularis	PV	—	12·7	12·7	53·1	621	30·3	—	—	
	SW	—	12·3	12·7	61·7	627	31·2	—	—	
ECHINOIDEA										
Strongylocentrotus droebachiensis	PV	420	9·7	9·6	48·6	488	29·0	7·2	6·00	152
Echinarachnius parma	PV	418	9·2	9·4	49·3	488	29·8	7·0	2·24	
	SW	458	8·5	9·7	33·8	492	25·4	8·1	2·15	
HOLOTHUROIDEA										
Holothuria tubulosa	PV	—	13·5	13·7	59·7	649	31·5	—	—	58
	SW	—	12·3	12·7	61·7	627	31·2	—	—	
H. tubulosa	PV	548	11·9	12·0	64·5	636	32·8	—	—	666
	SW	543	11·6	11·7	63·0	633	32·6	—	—	
Cucumaria frondosa	PV	456	7·4	8·9	33·2	501	25·5	7·8	2·14	614, 152
	Amb.	475	7·8	8·9	31·0	494	25·4	7·75	—	
	SW	458	8·5	9·7	33·8	492	25·4	8·1	2·15	
Caudina chilensis	PV	439	11·3	9·7	53·9	528	27·6	—	—	411
	SW	435	10·5	9·7	54·3	529	27·0	—	—	
Chiridota laevis	PV	420	9·7	10·2	55·2	488	30·2	7·0	2·24	152
	SW	458	8·5	9·7	33·8	492	25·4	8·1	2·15	

Abbreviations: PV, perivisceral fluid; Amb., ambulacral fluid from the water vascular system; SW, sea water.

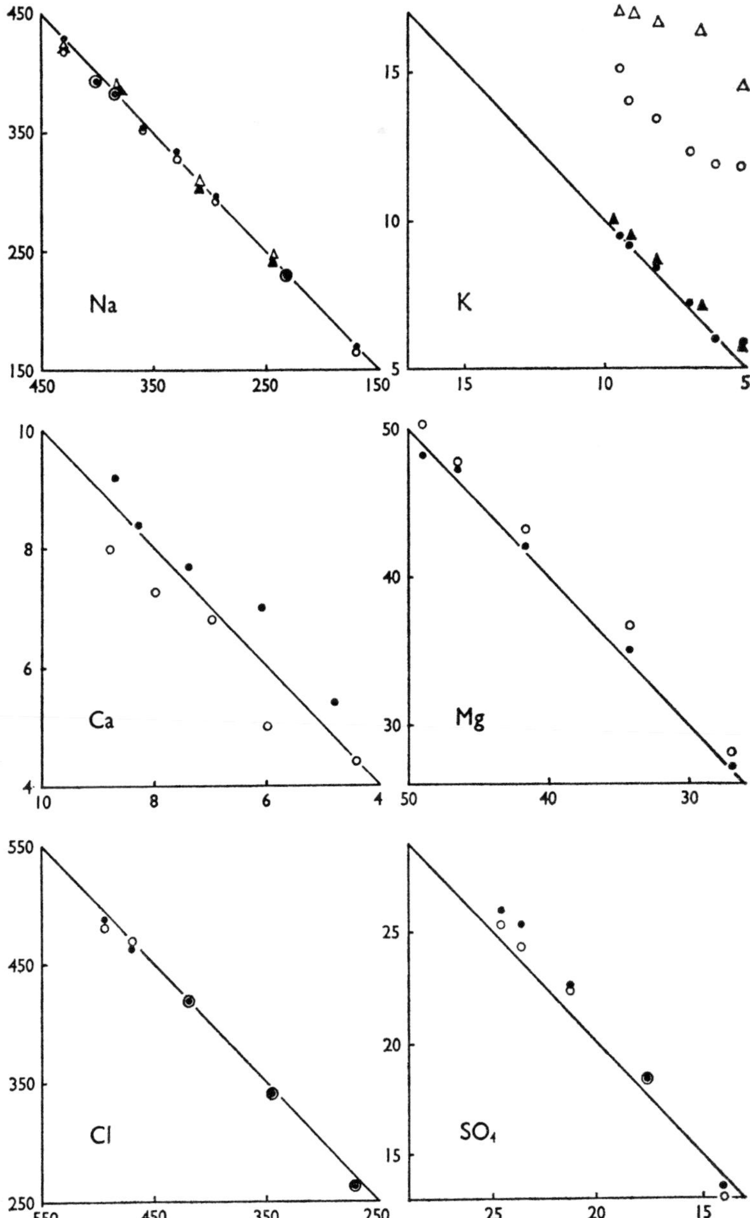

FIG. 5. Ionic regulation, in *Asterias rubens*, of sodium, potassium, calcium, magnesium, chloride and sulphate. Horizontal axes—external concentration in mM/l. Vertical axes—internal concentration in mM/l. Solid line represents isotonicity. All points are the average of ten determinations on different animals. Coelomic fluid: large animals ●, small animals ▲. Ambulacral fluid; large animals ○, small animals △.

and, like the asteroids, this is often accompanied by a fall in the magnesium level. The holothurians appear to have the weakest powers of ionic regulation, calcium and potassium often being marginally higher in the coelomic fluid than in the medium.

In animals with a hard calcified test, such as the echinoids, the accumulation of calcium is not surprising and the pH and carbon dioxide levels will in all probability have a direct influence upon such an accumulation. In the few species where these parameters have been estimated, both were slightly greater than in the environment, the fluid being more acid by as much as one pH unit. The buffering capacity of the body fluids has been investigated in *Solaster*. The ambulacral fluid had a greater capacity than did the perivisceral fluid.[288,689] From the equilibrium constants given in Harvey,[330] however, it can be calculated that there are actually fewer carbonate ions in the perivisceral fluid than in the sea water. The crystalline form of the skeletal calcium carbonate of echinoderms is calcite. The solubility product for this form is only about half attained in the perivisceral fluid, whereas the sea is virtually saturated or on occasions supersaturated with respect to this substance. It is possible, therefore, that this difference in the degree of saturation could allow for the uptake of calcium by the animal. This has been studied with particular reference to the larvae.[602,603] It is well known that in certain lamellibranchs small quantities of the calcareous shell pass into solution, acting as a buffer, during periods of anaerobic respiration whilst the animal is exposed by the tide.[211] Whether or not a similar function is performed by the echinoderm skeleton is not known, but the present author was unable to find any increase in the calcium concentration of the perivisceral fluid of starfish stranded by the tide which was not due to a general increase in salinity.

The other major coelomic compartment of echinoderms is the water vascular system. In both asteroids and echinoids an increase in the potassium concentration of between 20 and 90% compared with the external medium has been observed. In diluted media the asteroid *Asterias rubens* retains its potassium at an even higher level, but no figures from normally euryhaline starfish are available. Apart from this one example, the other major ions appear to be present in much the same concentration as they are in sea water. The quantity of protein present in the ambulacral fluid of this species is about 1·5 g/l, Such a quantity would be insufficient to retain the observed amount of potassium in a Donnan equilibrium. Experiments indicate that the potassium rapidly diffuses away when the ambulacral fluid is dialysed against sea water. Hence it would seem likely that this ion is present in the ambulacral fluid as a result of some form of active transport. The significance of this will be considered further in Chapter 8.

Organic Components of Body Fluids

The organic constituents of the coelomic fluids, although varied, are not present in sufficient quantities to affect the osmotic pressure, but they are of importance in maintaining the intracellular osmotic pressure and also have a nutritional role. The amino acid content of the perivisceral fluid has been estimated chiefly in this context and further details will be found in Chapter 1. The regulation of the concentration of such substances is most likely to be the result of the general metabolic activity of the animal rather than the operation of a specific transport mechanism. The protein of the perivisceral fluid has been variously estimated at between 0·2 g/l in *Holothuria tubulosa*[88] and 0·6 g/l in *Marthasterias glacialis*.[665] The male *Caudina chilensis* is reported to have a higher serum protein nitrogen and sulphur content than the female, but a slightly lower phosphorus content.[747] Reducing sugar levels also vary according to the nutritional state of the animal which depends upon such factors as the rate of hydrolysis of dietary polysaccharide and uptake of the products by the tissues. Little information is available regarding the lipids contained in the coelomic fluids. In the asteroid *Pisaster ochraceous* and thee chinoid *Strongylocentrotus purpuratus* values of 16·8 and 11·6 mg% respectively have been reported,[313] but the characterization of lipids, especially when only small quantities of starting material are available, is a notoriously difficult field. However, the presence of cholesterol in the perivisceral fluids of both *P. ochraceous* and *Picnopodia helianthoides* has been established at a concentration of 1 mg% and 4 mg% in *S. francescanus*.[560] The total lipid content of the whole animal and the significance of the occurrence of cholesterol in a group normally considered to contain only stellasterols, will be discussed in Chapter 5.

Permeability

Direct measurement of the permeability of the integument to ions seems to have been somewhat neglected and no modern work involving the use of radioactive isotopes appears to have been conducted. The rate of equilibrium of the iodide ion was followed in *Echinus esculentus* when sodium iodide was added to the water in which they were living.[47] No significant differences in the rates of penetration were observed when the mouth, anus or other regions of the integument were sealed. It was therefore concluded that the general body surface was extremely permeable to this ion. Koizumi,[413] working with isolated portions of the body wall of *C. chilensis*, found that potassium penetrated most rapidly, followed by sodium, calcium and magnesium in that order. No difference was found upon reversal of the direction of ion flow. However, the potassium content of the perivisceral fluid of this animal is slightly above that of the sea water in which it lives and it seems likely that the body wall is not acting as a passive molecular sieve but must be actively transporting some ions. Some sort of secretory activity has already been attributed to the integument in connection with the loss of ammonia from the perivisceral fluid.[194] Using lithium ions at a concentration well below their toxic level, there was no difference between influx into the perivisceral fluid or ambulacral fluid. Similar rates of efflux from both these cavities was also obtained, results which, like those of Koizumi, would seem to rule out any differential permeability.[61]

Skeletal Components

The gross chemical composition of echinoderms has been intensively investigated for a number of reasons. Vinogradov[797] provides an exhaustive account of the inorganic components. Water makes up between 65 and 80% of the total weight of echinoderms, echinoids containing the least water. Of the remaining dry material, about half is skeletal in origin and the rest organic. The skeleton consists of between 90 to 95% of calcium carbonate in the form of calcite and the rest is chiefly magnesium carbonate. Strontium is also present in trace quantities and in *Dendraster* the calcium/strontium ratio appears to vary inversely with the temperature of the water[628] and manganese is similarly affected.[324] The composition of the skeleton is not constant within the animal and it has been shown that in echinoids the inter-ambulacral plates, chewing apparatus and the short spines have a higher percentage of magnesium than does the rest of the skeleton. Warm water species have a higher magnesium concentration than do those from colder waters. Apparent exceptions to this rule, such as certain echinoids from the Peruvian coast, were in fact found to be due to the cold water brought up by the Humboldt Current.[138] Though it may take the form of dolomite or magnesite, the actual crystal form of the magnesium carbonate is uncertain.

Each skeletal unit is most probably a single crystal with a warped molecular lattice where necessary to permit curvature of the test or skeleton. Much crystallographic data has been amassed chiefly for morphological or palaeontological purposes. In a series of papers, Raup has shown that the majority of echinoids have the *c*-axis either perpendicular or tangential to the test with very little variation between the oral and aboral extremities.[653–656] Those species possessing a perpendicular arrangement generally have a greater degree of curvature.

In addition to those elements already mentioned, a number of others have also been recorded in trace quantities including barium, iron, aluminium, copper, zinc, manganese, cobalt, nickel, boron, cadmium and vanadium. This latter element occurs in the corpuscles of certain species of holothurian at a concentration many times that of the environment and poses interesting problems regarding the accumulatory mechanism reminiscent of the now classical situation in certain ascidians (see Chapter 7). In addition to carbonates, the skeleton also contains the sulphates and phosphates of the alkaline earth metals. A dietary relationship between the level of iron in the forage plants of the echinoid *Tripneustes esculentus* from two different localities and the level in the body has recently been established.[737]

The inorganic composition of the non-skeletal portions of echinoderms is poorly known. Only the holothurian *C. chilensis* has been investigated in any detail by Koizumi. The values obtained for the plasma, corpuscles, skin and longitudinal muscle are shown in Table 4 together with the levels found in the muscle of *Thyone briareus*. Initially it would appear that the holothurian muscle has a much lower sodium/potassium ratio than does vertebrate muscle and this aspect together with more recent work upon the partition of ions will be considered in

TABLE 4. INORGANIC COMPOSITION OF VARIOUS COMPONENTS OF HOLOTHURIANS

Species and component	mM/kg fresh weight						Reference
	Na	K	Ca	Mg	Cl	SO_4	
Caudina chilensis							
plasma	460	11·8	10·7	50·5	523	29	
corpuscles	202	176	6·9	14·5	126	131·5	
skin	302	78·7	270	90·4	277·5	95·5	
muscle	191	138·5	88·7	39·2	122	65·2	411–415
Thyone briareus							
muscle	173	169	14·0	—	204	—	735

Chapter 11. During the course of an investigation of the ionic regulatory powers of *A. rubens* the present author estimated the sodium and potassium concentration in the gonads over a period of one year. The results for males and females were recorded separately and yield an interesting trend. On the Kent coast, where this work was carried out, the animals spawn in late March or early April, the gonad weight as a percentage of the total body weight then being about 12%. In the males the sodium concentration was minimal in the spring at about 95 mM/kg wet weight, rising to about 275 mM/kg wet weight in the autumn. The potassium concentration on the other hand followed an inverse pattern, being maximal in the spring at about 125 mM/kg wet weight, falling to about 75 mM/kg wet weight in the autumn. In the females the pattern was similar but less pronounced. These changes, shown in Figs. 6 and 7, are obviously associated with the histological and cytological changes occurring in the gonad during the course of differentiation and maturation. The ionic composition of the eggs is similar to that of the intact ovary when mature.

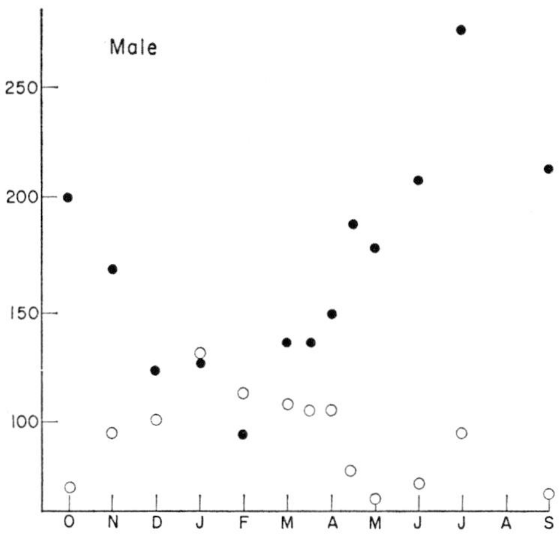

FIG. 6. Annual fluctuation in the ionic composition of the gonads of *Asterias rubens*. Male. Horizontal axis = month of the year. Vertical axis = concentration in mM/kg wet weight. Solid circles = sodium; hollow circles = potassium.

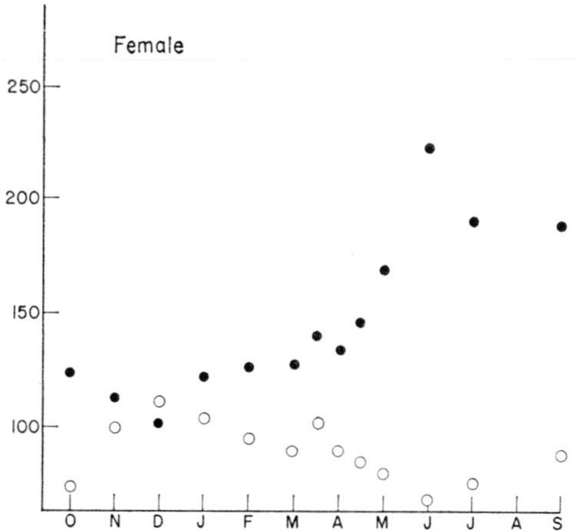

FIG. 7. Annual fluctuation in the ionic composition of the gonads of *Asterias rubens*. Female. Horizontal axis = month of the year. Vertical axis = concentration in mM/kg wet weight. Solid circles = sodium; hollow circles = potassium.

Use as Fertilizers

The high calcium content of asteroids coupled with their relative abundance in certain parts of the world has suggested to a number of workers their possible use as a fertilizer or as an animal feeding stuff. A thorough investigation of this possibility has been carried out[284,461] and yielded much interesting data upon the growth of various animals, including chickens, when starfish meal is added to their diet. The composition of such meal obviously varies, but Galtsoff and Loosanoff quote the following as typical:[284]

	% wet weight		% wet weight
Albumen	31·6	Ash	34·9
Fat	6·9	Sand	3·9
Water	12·1		

However, nutritionally, the protein appears to be of low grade, growth rates being relatively poor, and the use of this meal does not appear to be justifiable either upon economic or physiological grounds. Extracts of starfish have been shown to give rise to pathological conditions, such as bloat, when given to sheep, chicks and fish[331] and it is possible that the presence of these toxic principles was the reason for the poor growth rates in these feeding experiments. It is a widespread belief amongst fishermen that lightly cooked starfish are toxic when fed to cats and dogs. However, not all terrestrial vertebrates seem to be so affected, for herring gulls seem to feed voraciously upon starfish without ill-effect.[755] These birds also feed upon echinoids, cracking the tests by dropping the animal on to exposed flats, then alighting to devour the gonads. Such behaviour is strongly reminiscent of the manner in which snails are dealt with by various other species of bird.

Organic Composition

The gross organic chemical analysis of whole animals into carbohydrate, protein and fat has received a considerable amount of attention. Early in this century Meyer[510] published a number of such

TABLE 5. GROSS ORGANIC COMPOSITION OF ECHINODERMS

Species	Composition as % dry weight			
	Ash	Carbohydrate	Protein	Fat
ASTEROIDEA				
Asterias glacialis	45·4	—	42·6	10·6
A. rubens	47·4	9·1	34·9	5·7
A. rubens (North Sea)	38·0	19·4	32·8	12·9
A. rubens (Baltic Sea)	34·9	17·2	34·9	12·4
Hippasterias sp.	48·4	—	49·1	5·2
ECHINOIDEA				
Echinus sp.	83·2	7·1	8·4	0·68
Spatangus sp.	78·5	6·6	9·1	1·62
HOLOTHUROIDEA				
Cucumaria sp.	17·4	27·1	49·7	5·7

analyses and his results with those of Schlieper,[703] recalculated where necessary, are set out in Table 5. Recently Giese and his co-workers have devoted much attention to American West coast asteroids and echinoids and their results are reviewed by Giese.[296] Whilst information of this type is obviously of importance, it is greatly diminished in some cases by lack of data upon relevant conditions such as water temperature, season and size of the animal. It is well known that many echinoderms will store large quantities of fat and glycogen especially when approaching the breeding season, and this aspect is considered in Chapter 10.

Regrettably little is known of the exact chemical nature of these substances. Much of the carbohydrate fraction is probably glycogen, but the presence of complex glyco- and muco-proteins have been reported.[506,557b] The protein components are obviously extremely varied, but one particular protein has been intensively studied, namely haemoglobin, and this is considered in Chapter 5. The lipid fraction may vary in abundance from 0·7% dry weight in *Echinus* to 10% in *A. glacialis*. Assuming an average water content of some 70%, a 100 g animal will contain from 0·2 to 3·0 g of fat. This fat is divisible into two fractions, the saponifiable and the unsaponifiable, this latter being considered in Chapter 5. A long series of papers which originally appeared in the *Journal of the Chemical Society of Japan* dealing with echinoderm oils[763–766] has been reviewed by Toyama *et al.*[767] Oils extracted from the ophiuroids *Gorgonocephalus caryi* and *Ophioplocus japonicus* contained a variety of saturated and unsaturated fatty acids including palmitic, oleic and some highly unsaturated ones such as clupanodonic acid. Extracts of salted ovaries of the echinoids *S. pulcherrimus* and *Heliocidaris crassispina* yielded many saturated fatty acids which included myristic, palmitic, stearic plus some arachidic and behenic acid. The unsaturated fatty acids were chiefly zoomaric, oleic, the more unsaturated eicosaenoic and docosaenoic acids. Investigations upon *Pisaster ochraceous* have yielded a similar picture.[673] Some 18 to 29% of the fatty acids consist of arachidonic acid ($C_{19}H_{31}COOH$, 5.8.11.14 eicosatetraenoic acid). This substance has been shown to be effective in curing deficiency symptoms and promoting growth in fat-deficient rats. Most of the tissues of *Pisaster* had a high proportion of the C_{20} and C_{22} unsaturated acids, except, surprisingly, the gut caeca. Such polyunsaturated acids would account for the high iodine number found for the fat of *Asterias*.[350]

Various enigmatic compounds have been isolated from echinoderms to which no specific function can as yet be ascribed. For example, in 1963 Ackermann and Hoppe-Seyler[6] isolated a cyclohexanolmonoethyl ether from *A. rubens* with an empirical formula $C_7H_{14}O_6$. They named this substance Asterite and noted that it was the first time that one of this group of substances had been isolated from the animal kingdom. More recently Quin recorded 2-amino ethyl phosphoric acid from the related *A. forbesi*,[647] a substance which although of chemical interest in possessing a carbon–phosphorus bond is biologically baffling.

CHAPTER 5

BIOCHEMICAL AFFINITIES

ZOOLOGICAL studies are concerned not only with the detailed examination of individual species in isolation, but also with their relationship to the other members of the animal kingdom. There is a long history of classical comparative anatomical investigations, spurred on since the middle of the last century by the publication of Darwin's *Origin of Species*. Such investigations have reached the point where an authoritative account of the development and function of a wide range of structures can be given. Hence it is possible to assess, with a fair degree of accuracy, their evolutionary history and phylogenetic significance. Biochemical studies performed with an evolutionary viewpoint are much more recent. In many cases an exact knowledge of the function of the substances under consideration is not yet available and often there is a relative paucity of information upon their distribution. Even the characterization of the compounds has sometimes been inaccurate. Some caution, therefore, would not seem out of place in cases where biochemical evidence is at present in dispute with classical phylogeny. The late Professor Bergmann, one of the foremost workers in the field of the comparative biochemistry of steroids, said[48] (1949): "It is dangerous and frequently misleading to base significant conclusions concerning comparative biochemistry on data derived from but a few representatives of a given phylum." To recast a phylum which as long ago as 1900 was said by Bather[44] to be "one of the best characterized and most distinct phyla of the animal kingdom" upon the evidence of the occurrence in a relatively few species of only some compounds, whose function is not always clear would seem therefore somewhat premature.

α-Glyceryl Ethers

So far, two main groups of substances have been employed in this context, namely the sterols and the phosphagens, and it should be noted that the phylogeny deduced from their separate study is not always strictly complementary. However, one minor group of substances will be considered first of all. Echinoderms contain about twenty times as much unsaponifiable fat as do vertebrates and from this fraction has been isolated the sterols and a white crystalline material shown to contain the substance batyl alcohol or occasionally some of its homologues. Batyl alcohol is a glyceryl ether with the formula $CH_2OH.CHOH.CH_2O(CH_2)_{17}.CH_3$ and was first isolated from asteroids and holothurians in 1943.[500] Its supposed absence from echinoids has provided some of the biochemical evidence for proposing a divergence in the echinoderm stock into asteroids and holothurians on the one hand and echinoids and ophiuroids on the other.[362] Such a dichotomy is quite distinct from that usually presented which would hold that the Asteroidea and Ophiuroidea are more closely related and indeed are sometimes combined in a super class Stellaroidea. The echinoids *Strongylocentrotus pulcherrimus* and *Heliocidaris crassispina* contain not only batyl alcohol itself but also homologues chimyl and selachyl alcohols.[767] Latterly, other invertebrates such as the mollusc *Mytilus* and various tissues from vertebrates such as bone marrow and liver have also been shown to contain batyl alcohol.[760] Although nothing is known of the function of batyl alcohol in invertebrates, in the mammalian body it seems to be associated with erythropoiesis and has even been suggested as a treatment for agranulocytosis. Its occurrence in echinoderms, a phylum where amoebocytes play a significant role in the animal's economy, is of great interest. The build-up of α-glyceryl ethers in echinoderms is thought to be due to the absence of the appropriate splitting enzyme. Isolated starfish diverticula incorporate C^{14} acetate and glycerol into α-glyceryl ethers at a rate comparable with other lipids. They do not appear to be synthesized in response to low oxygen tension as might be expected.[385] It would therefore seem reasonable to infer that phylogenetic significance should no longer be attributed to the distribution of this substance.

Steroids

A more important substance, phylogenetically speaking, in the unsaponifiable fat fraction is the steroids. A vast amount of work in the field of steroid chemistry and distribution has been carried out by Bergmann and his associates whose long series of papers culminated in two reviews in 1949 and 1963[48,49]. In Table 6 can be found the principal results of their work. The history of invertebrate sterols goes back some way. In 1909 Doree[203] noted the peculiarity of sterols from invertebrates and later Kossel and Edelbacher named the steroid from the starfish *Asterias rubens* Stellasterol.[420] Bergmann and Stansbury later added the monounsaturated stellastenol to the list.[50,51] With improving techniques more sterols were isolated and characterized by various groups of workers and as a result a number of synonomies have occurred. For example, one component named astrol by Kossel and Edelbacher was later called asteriasterol and is now known not to be a sterol at all but is identified as batyl alcohol. Hitodestrol from *A. amurensis* would appear to be synonymous with Pateriasterol from *Asterina pectinifera* and also α-spinasterol isolated from spinach.

TABLE 6. ECHINODERM STEROLS

Species	% Unsap. fat	Principal sterol
ASTREOIDEA		
Astropecten scoparius	12	Hitodestrol
Asterina pectinifera	20	Hitodestrol
Asterias rollestoni	10	Hitodestrol
A. rubens	39	Stellasterols
A. forbesi	34	Stellasterols
ECHINOIDEA		
Heliocidaris crassidus	—	Cholesterol
Centrechinus antillarum	20	Cholesterol
Lytechinus variegatus	—	Cholesterol
Tripneustes esculentus	26	Cholesterol
Arbacia punctulata	22	Cholesterol
HOLOTHUROIDEA		
Holothuria princeps	11	Stellasterols
Cucumaria chronjhelmi	11	Stellasterols
OPHIUROIDEA		
Ophiopholis aculeata	—	Cholesterol type
CRINOIDEA		
3 spp (unnamed)	—	Cholesterol type

The picture of the distribution of sterols which had emerged by the early 1950s provided some grounds for believing in a basic dichotomy of the phylum. The asteroids and the holothurians on the one hand seemed to possess the stellasterol complex and the echinoids and the ophiuroids on the other hand seemed to possess a sterol in common with the vertebrates, namely cholesterol. However, notwithstanding Bergmann's rider, data drawn from only sixteen species spread over the five

TABLE 7

Species	Sterol	Reference
CRINOIDEA		
Comanthus japonica	mixture of Δ^5 sterols, pro-vitamin D	499
Comatula sp.	crinosterol (Δ^5) 24β methyl 22 dehydrocholesterol	71
ECHINOIDEA		
Strongylocentrotus pulcherrimus	Δ^5 cholesterol pro-vitamin D	767
Heliocidaris crassispina	Δ^5 cholesterol pro-vitamin D	767
OPHIUROIDEA		
Gorgonocephalus caryi	Δ^5 β-sitosterol	767
Ophioplocus japonicus	Δ^5 β-sitosterol	767
Ophiopholis aculeata	Δ^5 β-sitosterol, γ-sitosterol, poriferasterol, stigmasterol	499
ASTEROIDEA		
Luidia quinaria	Δ^7-spinastenol + others	763
Asterina pectinifera	α-spinasterol, Δ^7-spinastenol, Δ^7-cholestenol	766
A. pectinifera	22·23 ergostenol (=patiriasterol)	442
Asterias amurensis (=*A. rollestoni*)	Δ^7-cholestenol	764
A. amurensis	Δ^7-stigmastenol	500b
Coscinasterias acutespina	Δ^7-sterol, unnamed	767
HOLOTHUROIDEA		
Cucumaria chronjhelmi	Δ^7-cholestenol	499

classes were cited as providing more substantial support for the union of the Echinoidea and Ophiuroidea than did more classical studies.[362] This proposal has met with considerable discussion and the reader is referred to papers by Kerkut,[400b] Nichols,[579] Fell[238] and Bolker.[70]

Latterly, interest in the invertebrate sterols has centred upon a number of Japanese workers whose results are summarized in Table 7. Attention has been drawn to the fact that, regardless of the nature of the side chains and their degree of saturation, those asteroids and holothurians so far investigated possess sterols with the Δ^7 configuration whilst echinoids, ophiuroids and crinoids possess sterols with the Δ^5 configuration.[767] In this instance the division relies upon differences in a more fundamental part of the molecule, namely the steroid nucleus, which is less labile than the side chains. In the case of invertebrate sterols generally, just as with the enzymes associated with nitrogenous excretion, evolution appears to have been associated with a reduction of diversity. Whereas the simpler animals have a wide range of sterols, the more advanced ones have fewer examples, culminating in the sole use of cholesterol by vertebrates. The possession by living echinoderms of a variety of sterols may hence represent nothing more than a differential survival from an even more extensive array of sterols which may have existed in ancestral forms. Metabolic pathways and their products must necessarily be a reflection of the genetic makeup of the organism; nevertheless, until more information is available upon the function of these sterols it would be unwise to place too great an emphasis upon the phylogenetic aspects of their distribution, which up to the present appears to have been oversimplified.

In the vertebrates, cholesterol appears to play a dual role. Together with lecithin, it is an important element in the construction of cellular membranes and it is an important starting material for the synthesis of bile acids, where it is conjugated with taurine or glycine. Closely allied are the steroid hormones and vitamin D, a substance associated with the process of ossification. With this background, a certain amount of work has been directed to isolation of other related steroids from echinoderms. No oestrogenic activity from ovaries of *Asterias forbesi*, *Arbacia punctulata* and *S. droebachiensis* could be detected,[316] but recently oestradiol-17-β has been extracted from the ovaries of the starfish *Pisaster ochraceous* and the sea urchin *S. franciscanus*, where activity varied from 0·04 to 10 μg/kg wet weight.[90] Some progesterone activity was also tentatively identified. Cornification of rat vaginae followed upon injection of ovarian extracts of *Lytechinus variegatus*, *Echinometra* spp. and *Stichopus moebii*.[201] If this hormone were synthesized *de novo*, then a different synthetic route from that found in vertebrates must have been followed, for cholesterol has not so far been found in asteroids. Cell suspensions of *A. punctulata* and *Echinarachnius parma* were able to convert oestradiol-17β into oestrone using DPN as a co-factor, testosterone and 4-androstenedione being converted at a lesser rate.[333] Conversion from a dietary precursor must not be ruled out, for working with *Pisaster*, the synthesis of 7:24 ergostadienol from 24-methylene-cholesterol found in the molluscs upon which the starfish feeds has been demonstrated.[226] Preparations of isolated gut tissue from *Strongylocentrotus franciscanus* can also sulphate oestradiol.[166] The data were consistent with the possibility that oestradiol is synthesized in the gut and can be excreted therefrom in a water soluble form. The appearance of pro-vitamin D in several of the above-mentioned animals is also of interest in view of the high degree of calcification found in echinoderms.

Phosphagens

In the early years of this century research upon the guanidine derivatives led on one occasion to the biochemical redefinition of the animal kingdom into creatinates and acreatinates.[441] However, as creatine had already been found in a number of invertebrate phyla,[560] this suggestion is really only of historical interest and the sporadic occurrence of creatine in the echinoderm classes (see Chapter 2) has little significance. However, speculations regarding phylogeny deduced from a study of the distribution of the phosphagens phosphocreatine (PC) and phosphoarginine (PA) is a little more soundly based. A number of surveys on the distribution of these substances and of their kinases—creatine phosphoryl transferase CPT, and arginine phosphoryl transferase—APT have been published.[35,36,218,552,572] In general the results follow a similar course to that for the parent guanine, namely that PC was characteristic of the vertebrates and PA characteristic of the invertebrates. However, the more recently published papers show up a picture analogous to that found for the sterols. Once again a single substance is characteristic of the vertebrates, PC, but in the invertebrates a number of phosphagens appear to co-exist including PA, PC, N-phosphoryltaurocyamine, N-phosphorylglycocyamine and N-phosphoryllombricine. Whilst the distribution of these latter substances has not been specifically investigated in echinoderms, the occurrence of both PA and PC is well documented. Table 8 summarizes data from 29 species, together with some information about chordates to illustrate the echinoderm–enteropneust–urochordate affinities. It will be seen that in one species of asteroid, *Asterina gibbosa*, neither phosphotransferase is present, whilst in a number of echinoids both PA and PC together with their kinases are present. In several species of echinoid, including *Anthocidaris crassispina*, the eggs contain only the arginine system whilst the sperm contain the creatinine system together with very small amounts of PA.[823–826] In general, therefore, it may be said that representatives of four of the classes of echinoderm so far examined possess the phosphoarginine system and the ophiuroids possess only the phosphocreatine system, whilst some echinoids possess both. Such evidence which is still in a rather incomplete form cannot be used soundly to lend support for the closer union of ophiuroids and echinoids.

Creatine is synthesized by the methylation of glycocyamine, itself a substance capable of functioning as a phosphagen when phosphorylated as for example in *Nereis diversicolor*[753b] and more recently in the echinoderm *Henricia sanguinolenta*.[552] Glycocyamine is formed from arginine and glycine by a transamidination reaction. Thus each one of the three guanidines in this synthetic pathway could function independently as a phosphagen if the necessary phosphorylating enzymes were present. Whilst N-phosphoryllombricine is a typically annelid phosphagen, its precursor, serine ethanolamine phosphodiester (SEP), has now been recorded from five unnamed species of echinoderm.[219] Another substance, closely related to a phosphagen, is N,N-dimethyltaurocyamine. Under the name of Asterubin it was isolated from *Asterias rubens* and *Marthasterias glacialis* by Ackermann.[5]

Phosphagens are small, relatively labile, molecules and as such may be subjected to metabolic pressures not necessarily closely associated with their main function. It would seem reasonable to suppose, there-

TABLE 8. SOME ECHINODERM PHOSPHAGENS AND PHOSPHOTRANSFERASES

Species	Phosphagen and enzyme			
	PA	APT	PC	CPT
CRINOIDEA				
Antedon bifida (whole)	+	+		
A. mediterranea (whole)	+			
HOLOTHUROIDEA				
Holothuria tubulosa	+	+		
H. forsakali (long. muscle)		+		
Synapta inhaerens (b.w.)	+			
Stichopus moebii		+		
ASTEROIDEA				
Marthasterias glacialis (tube feet)	+			
Asterias rubens (tube feet)		+		
A. forbesi	+			
Astropecten irregularis		+		
ECHINOIDEA				
Arbacia punctulata (jaw muscle)	+	+		
Centrostephanus rogersii	+			
Diadema setosum		+		
Strongylocentrotus lividus (jaw muscle)	+		+	
Paracentrotus lividus	+	+		+
Echinus esculentus (jaw muscle)	+	+	+	+
Heliocidaris erythrogramma	+	+	+	+
Lytechinus variegatus		+		+
Anthocidaris crassispina				
sperm	+		+	+
egg	+	+	−	−
OPHIUROIDEA				
Ophioderma brevispina (arms)			+	
Ophiothrix fragilis (whole)			+	+
Ophiocomina nigra (arms)				+
Ophionereis reticulata				+
HEMICHORDATA				
Balanoglossus clavigerus			+	
UROCHORDATA				
Styella mammiculata		+		+
Ascidia mentula	+		+	
Pyura stolonifera			+	+
Amphioxus				+
Vertebrates			+	+

fore, that the enzymes themselves, being proteins, and hence the primary products of gene action, would give a more reliable reflection of phylogeny rather than the smaller molecule. Such a thesis has served to direct the attention of Watts and his co-workers towards such enzymes.[552,803] Their measurements show that echinoderm arginine kinase has a molecular weight about twice that of other similar invertebrate enzymes, namely about 800,000 and is a dimer. The molecular weight of creatine kinase suggests that it is a tetramer. On a variety of evidence, which includes the nature of the catalytic site, its pH dependence, and amino acid composition, it is suggested that these enzymes are homologous within the biochemical meaning of the term.[803] Wald has made the suggestion that arginine kinase (APT) is the more primitive enzyme although its evolution to the creatine kinase (CPT) is probably more complex than the single mutation which he proposed.[800] Polymerization of the arginine kinase sub-unit to a dimer may have then been followed by a mutation so that the enzyme could then bind creatine. Alternatively the mutation may have occurred first and the ability to dimerize may then have evolved independently in both enzymes. In the latter case, creatine kinases with a molecular weight of 40,000 should exist, but so far only the tetramer has been been recorded from echinoderms.

The phylogenetic importance of the distribution of phosphagens is no longer so clear-cut as was at one time thought. One of the difficulties in interpreting biochemical evidence of this nature, especially when related to the higher systematic taxa, is the absence of data from ancestral types, evidence which the morphologist usually has from the fossil record. Information from living representatives of ancestral groups does not necessarily reflect the ancestral condition. Thus in the echinoids PC occurs only in association with PA and then only in the most advanced Camerodonta, all the other groups containing PA only. The exact role of these two co-existing phosphagens is not clear. Two species of Asteroidea have now been recorded as containing CPT, one of them, *H. sanguinolenta*, also containing glycocyamine phosphoryltransferase. The division of phosphagens between the gametes has already been referred to and now the curious instance of *Psammechinus miliaris* has also been recorded.[552] Here again the eggs contain only the arginine system, but the lantern muscle from the adult contains only the creatine system. It is postulated that the use of the creatine system by the sperm would confer a selective advantage in that the use of a phosphagen which is not a metabolite could obviate the drain on arginine in nuclear histone synthesis. Such potential may or may not be suppressed by genetic mechanisms in later development. The possession of the creatine system may therefore only represent common ancestry and not close kinship as has been suggested by several authorities.

Proteins including Haemoglobin

A dictum, usually ascribed to Brachet is "DNA makes RNA and RNA makes Protein". The only proteins so far considered in echinoderm relationships have been those concerned with phosphagen metabolism, the structure of the nucleic acids having received little attention. The behaviour of deoxyribonuclease from *S. droebachiensis* in the presence of various salts was, not surprisingly, unlike that of enzymes extracted from an annelid, a mollusc and a crustacean.[683] It more closely resembled the vertebrate or Type II DNAase, for which point the authors claim a phylogenetic significance. If, as we believe, the proteins are closer to the genes than the diverse products of enzymic activity, then similarity of the protein component as measured by immunochemical techniques should prove of interest, although the same strictures regarding common ancestry must also be applied here. Kubo[426] prepared antisera in rabbits of five species of asteroid. Precipitin reactions coupled with total nitrogen determinations as a means of quantizing the results suggested that the Phanerozonia were a heterogeneous assemblage. On the basis of the salting out of the proteins from all five classes of echinoderm, it was concluded that the crinoids, echinoids and ophiuroids were more closely related to each other than to the asteroids and holothurians, these latter forming a separate grouping.[611]

Finally the protein haemoglobin, found in a number of holothurians, has been examined by Manwell and his associates.[493,494] The holothurian *Thyonella gemmata* exists in two distinct morphological forms, termed "stouts" and "thins". Such morphological types also have distinct haemoglobin components, the haemoglobin from the "thins" being resolvable into eight components and that of the "stouts" into only five. This is due to the varying proportions of haemoglobin chains of differing lengths. In the "stouts" these consist of Type 1, subdivided into a, b, c, d, all containing the A chain, and Type 2, containing only the B chain. Additionally the "thins" possess Type 3 (A and C chains), Type 4 (B and C chains) and Type 5 (C chain only). In addition there seems to be some difference between the haemoglobin found in the perivisceral coelom and the water vascular system of the "thins", the former having a higher proportion of Type 3 and the latter a higher proportion of Type 2. Attention was drawn by this author to an analogous position found in the isoenzyme of lactic dehydrogenase where a mixture of five types occurs and the proportions vary during the course of development. Manwell points out[492] that the haemoglobin of the "thin" variety bears some resemblances to that of *Myxine* however, the remaining living vertebrates seem to bear greater similarity to the petromyzonts rather than the hagfishes.

CHAPTER 6

TOXINS AND IMMUNOLOGY

THE subject of echinoderm toxins has already been broached in connection with the feeding mechanisms of asteroids (Chapter 1). The proven ability of a starfish to overcome the resistance of its molluscan prey by seemingly purely mechanical force in no way completely rules out the possibility of the participation, even if only sporadically, of a toxic secretion. Extracts of starfish stomach have been shown to have either toxic or some other pharmacological action.[571,580,693] Cellular disintegration and discharge of nematocysts occurred in *Hydra* and *Aurelia* following the application of such extracts, such behaviour resembling that of immersion in a calcium-free medium. Some experimenters have, however, employed extracts of whole animals and thus it is difficult to be certain of the organ responsible for the production of the active substance. It is with those toxins not thought to be associated with the stomach that the bulk of this chapter is concerned.

Mucus

The possible presence of a toxin in the mucus secreted by the tube feet of crinoids during the process of feeding has already been mentioned, but the findings of Reichensperger[658] have not been reinvestigated. The mucus produced by various species has been investigated from time to time, but little is known of its chemical structure. Fontaine[254] suggested from his observations upon the metachromatic properties, that the mucus produced by the multicellular glands on the body of *Ophiocomina nigra* was a highly sulphated mucopolysaccharide of the heparin type. Vast quantities of this mucus may be produced in response to alarm conditions.[259] In some similar species of ophiuroid these mucus glands may also be the source of phosphorescent material.

Behavioural Responses

Other epidermal secretions, whilst not having a toxic effect in the narrow sense, can certainly trigger off quite marked behavioural responses in other animals. As a phylum, the molluscs seem to have a well-developed sense of chemoreception (see, for example, Kohn, 1961[410]), a sense which seems particularly attuned in the gastropods to the presence of secretions produced by some of their predators, namely echinoderms. Earlier work, reviewed by Bullock,[98] has been supplemented in recent years by many other workers, who have described essentially similar phenomena. Feder,[235] for example, notes that several gastropods show escape responses to *Pisaster ochraceous* and *Pycnopodia helianthoides* either upon direct contact or exposure to water in which these starfish had been living. It is claimed that the active principle is associated with the epidermal covering of the water vascular system, and later a yellowish substance was extracted from the tube foot epidermis which absorbed strongly on to colloids and was extremely potent in eliciting the flight reactions of *Acmaea* and other molluscs.[236] The snail *Tegula* fled from contact with these starfishes, which normally preyed upon it, escape reactions also being elicited by substances diffusing in the water.[827] Similar behaviour by *Diadora aspersa* to seven species of forcipulate asteroid has also been recorded, again direct contact being unnecessary for the characteristic flight reaction to occur.[495] This reaction consists of extension of the mantle over the shell to prevent adhesion by the tube feet. The raising of the shell in the water followed by alternate twisting movements is also resorted to as in the limpet *Acmaea*. A violent flipping movement is also recorded in *Nassarius vibex* upon contact with predatory starfish. Echinoderms will also exhibit flight reactions to other echinoderms. Prouho[646b] records that the starfish *Marthasterias glacialis* will move away from the urchin *Paracentrotus lividus* when bitten by its globiferous pedicellariae. An interesting comparison has been made between the behaviour of *Strongylocentrotus droebachiensis* and *Psammechinus miliaris* to this same starfish.[372] Whereas the latter species will defend itself with its pedicellariae, made all the more effective by the tangential bending away of the spines, the former urchin will only behave in this way if it is actually cornered, otherwise it fled before its attacker. Defensive reaction of the spines by their turning towards the attacker in the case of *S. purpuratus* was not recorded here. *Dendraster excentricus* will burrow into the substratum up to 2 ft in front of an advancing *Pisaster brevispinus*, a species which normally feeds upon it.[565] Vast quantities of mucus can be produced by the starfish *Pteraster tesselatus*, and gastropods, crabs and a sea cucumber placed in water containing such mucus rapidly die.[674]

Pedicellarial Toxin

An obvious epidermal structure much concerned in this phylum with offence and defence is the pedicellaria. Extracts of all types of pedicellariae from four species of echinoid were found, upon injection into a wide range of animals, including crabs, octopus, fish, lizard and rabbit, to be toxic and in all cases death was due to respiratory failure.[339] All the other species tested, including echinoids, holothurians and the frog, were apparently immune. Transfer of immunity from the frog to the crab was claimed to be possible by injecting into the crab 1 ml of frog serum. The general effect of the toxin was to produce a paralysis, the substance being thermostable. More recently Peres[623] has, with some reservations, confirmed these observations using only one of the species which had been earlier employed, namely *Sphaerechinus granularis*. He found that in addition to the globiferous pedicellariae, extracts of the ophiocephalus and tridactyle pedicellariae, axial organ, ambulacral and coelomic fluids were also toxic, although five to six times less potent. In the earlier stages of paralysis of several species of crustacean, the respiratory musculature seemed more resistant than other muscles, but eventually these too were affected and death ensued. The immunity of the frog to this toxin is only relative, a somewhat stronger dose than might have been expected being required to produce the typical symptoms. He further draws attention to the reported similarities with the toxin of coelenterate nematocysts, which appears to consist of hydroxyindoles resembling 5-hydroxytryptamine and N-dimethyl 5-hydroxytryptamine adsorbed upon mucoprotein.

Probably the best known examples of toxic pedicellariae are to be found in the echinoid genus *Toxopneustes*. These pedicellariae are

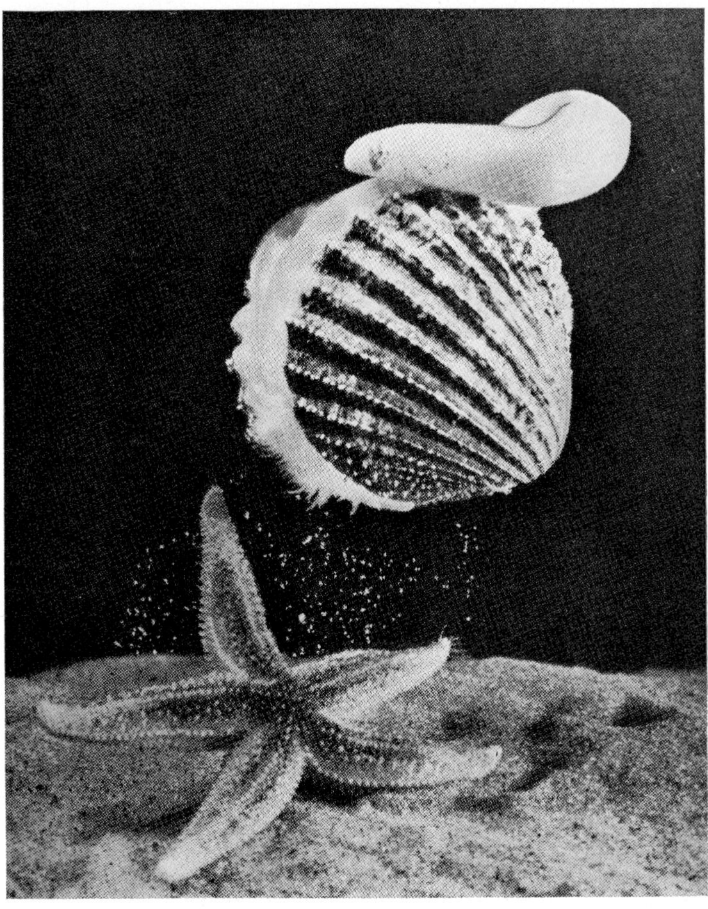

FIG. 8. The cockle *Cardium echinatum* in the middle of an escape jump elicited by the starfish *Asterias rubens*.

(*facing page 70*)

normally open or will open in response to slight mechanical stimulation, especially upon the sensory hillocks. Chemical stimulation results in immediate closure usually followed by emission of the toxin. In 1935 Fujiwara[283] reported that pain, dizziness, slight facial paralysis and impeded respiration resulted from the bites of several of these pedicellariae upon the finger. More recently the toxin of the giant pedicellariae of the same species has been more intensively investigated by Okada and his colleagues.[600,601] They showed that the white mucus was in fact a matrix in which were suspended dumb-bell-shaped corpuscles. Injection of the matrix into the oyster heart resulted in an increased tonus and a reduced amplititude of contraction. After some time, the preparation returned to a more normal type of behaviour. Injection of the corpuscles resulted in the potentiation of the contractions. It is well known that, in general, molluscan hearts are inhibited by acetylcholine and stimulated by 5-hydroxytryptamine and the existence within a single secretion of two seemingly antagonistic components mimicking the effects of these substances is of very great interest. The possibility that acetylcholine or some similar substance is indeed one of the components, is further supported by other workers.[505] From homogenates of *Lytechinus variegatus* pedicellariae, an easily dialysable substance was isolated which had a cholinergic action upon isolated guinea pig ileum, rat uterus, canine blood pressure, longitudinal muscle of holothurian body wall and the sea urchin lantern protractor muscle. Its action upon the amphibian heart was blocked by atropine and potentiated by eserine and prostigmine. The closely related *Tripneustes gratilla* yielded non-dialysable toxin from the pedicillariae which also elicited prolonged contraction of the isolated guinea pig ileum.[237] Chemical evidence suggested that the intermediate release of histamine by the toxin was responsible for these effects. In view of these powerful pharmocological properties it would be of interest to know the effect of these toxins upon the natural prey of the sea urchin both in feeding and defence. The spines of several species of Echinothuridae, the bases of which are surrounded by a poison bag often brownish in colour, are also toxic. More recently, effects similar to those recorded by Fujiwara have been shown to follow stabbing by the spines of the asteroid *Acanthaster planci*.[41,632] These seemed to abate after application of pethidine hydrochloride and anti-histamine cream. This animal was also observed to emit a strange orange-pink substance when stimulated. Such behaviour obviously makes more difficult the important task of control of this species on coral reefs.

Holothurin

It has been known for some time that fish placed in an aquarium previously occupied by the holothurian *Actinopyga agassizi* died. This species of sea cucumber possesses cuvierian tubules which do not become sticky when ejected into sea water and extracts of them are toxic to fish. The classic exception to this situation is the famous pearl fish *Carapus* (=*Fierasfer*) *bermudiensis* which seems to live indefinitely in the cloaca and respiratory trees of this holothurian without ill-effect. It is, however, killed by more concentrated extracts of the organ. Extracts of this organ from the "cotton spinning" species appear to be harmless, but the body wall would seem to be an alternative source of the toxin.[586] Nigrelli and Jakowska[588] list numerous species of holothurians toxic to fish and additionally extracts of the body wall and viscera of three European holothurians—*Holothuria tubulosa*, *H. poli* and *H. impatiens*—have also been shown to be toxic to several species of fish at dilutions as low as 1:1,000,000.[22] The extracts of *H. impatiens* proved to be the most powerful, but the effects of the toxin were always similar and irreversible. When immersed in such solutions the fish soon began to make rapid movements of the operculum, became agitated and frequently surfaced. Such facts are well known to the natives of the Pacific islands who use macerated sea cucumbers as a means of stupefying fish in rock pools, such fish being non-toxic when eaten. The nature and morphology of the gills of these fish seem important in determining the rapidity with which the poison takes effect. The toxin from the pedicellariae, it should be recalled, does not appear to effect the respiratory musculature for some time. Frog tadpoles and the isolated ileum of the guinea pig are also adversely affected by this toxin which also exerts a strong haemolytic effect in mice, both *in vitro* and *in vivo*, where death resulted from massive haemorrhages.[586] Injection of holothurin, as the toxin is called, into mice with Krebs-2 ascites tumors, lead to an increase in survival time with an observable inactivation of the cells.[742,743] Some antibiotic properties are also suspected.

The application of holothurin to the frog sciatic nerve has uncovered another aspect of its toxic properties.[276,277] It markedly and irreversibly reduced the amplitude of the spike potential without effecting the speed of conduction. In this respect its neurotoxic action is reminiscent of that of veratrine. Its action upon the rat diaphragm preparation was to produce a contracture without completely abolishing its ability to execute a twitch upon direct electrical stimulation. As the drug-induced contracture dies away, so does the response to electrical stimulation. The response to both these modes of stimulation is irreversibly destroyed at concentrations higher than 1×10^{-4}M. Surprisingly, physostigmine at concentrations between 5×10^{-10}M and 5×10^{-8}M will prevent a major portion of the blocking action of holothurin.[273] Neostigmine is also effective, but it is 10^4

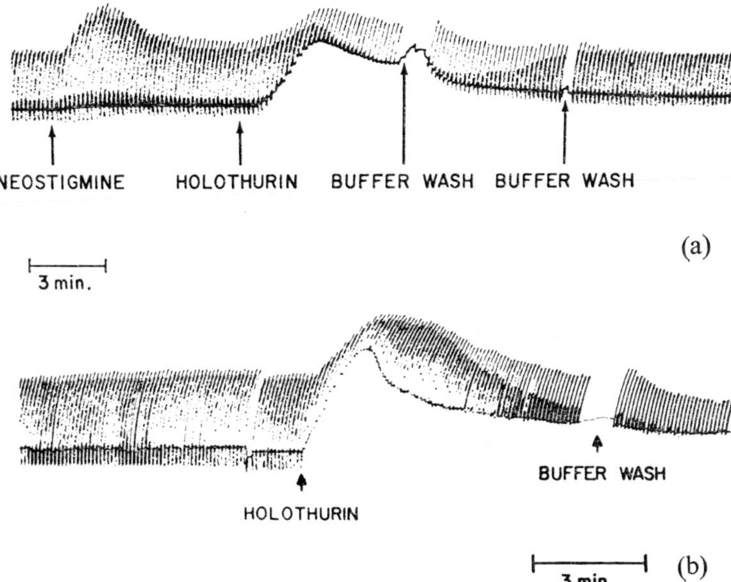

FIG. 9. Protective action of neostigmine. (a) Representative tracing of a rat phrenic nerve/diaphragm twitch responses protected by a $9{\cdot}9 \times 10^{-7}$M level of neostigmine against the normally irreversible destruction by a $1{\cdot}03 \times 10^{-4}$M level of holothurin. (b) Control tracing showing response to a $0{\cdot}99 \times 10^{-4}$M level of holothurin.

times less potent.[274] The protective action of physostigmine disappears at concentrations above 7×10^{-9}M and therefore bears no correspondence to the anticholinesterase activity of this substance. The possible mechanism for this interaction has been discussed[759] and special attention has been drawn to the axoplasmic changes in the node of Ranvier, and comparison made to the action of digitonin and Quillaja saponin. Embryological effects of holothurin include altering the gastrulation pattern of sea urchin embryos.[680] Extracts of four species of asteroid tube foot have an immobilizing effect upon *Arbacia* sperm and an animalizing effect upon the development of the larvae.[679] This consists of a hyperdevelopment of the apical tuft, thickening of the apical ectoderm, absence of the archenteron and suppression of the ento-mesodermal structures. Similar embryological effects are also produced by the mucus from certain fish and *Gonyaulax* toxin, but not by other steroid saponins such as ouabain and Quillaja toxin, which also includes antimetabolic activity such as retardation of the growth of head portions of planarians. For reviews of the earlier work upon the pharmacology of holothurin see refs. [132, 277, 586]

The chemical nature of holothurin has been intensively investigated. The possibility of its being a saponin was suggested as long ago as 1955 by Yamanouchi[820] who observed that the water in which sea cucumbers had died was often strongly foaming. The use of infra-red spectroscopy indicated the presence of a sulphuric acid residue attached to what was thought to be a steroid saponin.[129] The aglycone portion was named Holothurinogen A, though it is now called 22:25 oxidoholothurinogen, of which the deoxy analogue, 17:deoxy 22:25 oxidoholothurinogen, is also known. A second holothurin, designated holothurin B, which differs only in the sugar moiety, containing only D-quinovose and D-xylose, has recently been described.[828] It has now been established that holothurin A is a steroid saponin, the tetracyclic steroid nucleus being attached directly to the sulphuric acid residue. The sugar moiety consists of a linear chain of four monosaccharides—xylose, glucose, 3-oxymethyl glucose, quinovose (=glucomethylose) in that order.[131] In some respects this structure resembles the steroid alcohols scymnol and ramnol found in the bile of primitive vertebrates. The activity of the toxin appears to reside in two fractions; one, the cholesterol pre-cipitable fraction, contains the holothurin A. The loss of two sugar residues only moderately affects the activity, but loss of only half the esterified sulphate fraction results in an abrupt decrease of the blocking potency to zero.[275] The fact that saponins are only slowly absorbed through the gut wall may well be the reason why neither the fish caught by the method described above, nor dried sea cucumbers, trepang, are toxic when eaten.

Recent work has shown that the occurrence of steroid saponins is not peculiar to holothurians, although Burrage[103] found no evidence for the secretion of toxins by two species of asteroid and two species of ophiuroid. However, holothurin-like substances have been recorded from the asteroids *Pycnopodia helianthoides*, *Asterias forbesi*, *Patiria miniata*, *Pisaster ochraceous* and *P. brevispinus*.[661,662] The algycones may be different in each case, but the sugar moieties are similar. The absence of an α-β unsaturated lactone ring distinguishes these saponins from the plant cardiac glycosides. The toxins appear to alter the permeability of the red cell membrane, and the heart of the toadfish *Ospanus* was arrested in systole but could be restarted by elevation of the potassium level of the perfusing fluid. It was also highly toxic to *Fundulus* over a wide range of concentrations and appears to have the familiar neurotoxic effects and cytolytic properties. The starfish *A. forbesi* seems to be resistant to the toxins and would appear to have the interesting facility of being able to detoxify the solutions, possibly by enzymic activity upon the sugar moieties. A new steroid saponin recently isolated from the holothurian *Halodeima grisea* has been named griseogenin.[771] This substance would appear to contain the same steroid nucleus as holothurinogenin.

In conclusion it must be stressed that the properties of holothurin cannot be adequately described by comparison with any previously known substance. It is a more powerful haemolytic agent than saponin, but unlike saponin it stimulates haemopoiesis. Its neurotoxic effects are comparable to veratrine, but unlike veratrine it possesses an anti-metabolic action upon bacteria, protozoa and regenerating flatworms. It seems paradoxical that such a sluggish creature as the sea cucumber—one specimen in an aquarium was claimed to have remained stationary for more than two years[591]—should produce one of the most powerful biotoxins known. Its normal mode of employment by the animal is far from being understood. Studies on the physiology of commensalism between echinoderms and certain polynoid worms suggest that substances attractive to these worms are liberated into the water by *Evasterias troschelli* and *Stichopus californicus*.[187,188] Such substances disappeared from the aquarium within 24 hr and were species specific. It may be that minute quantities of highly toxic substances like holothurin could function by repelling other animals, not only predators, so keeping it free from sedentary creatures. It is singularly notable that as a phylum echinoderms are remarkably free from commencals and parasites. Cornman[157] has recently described what he believes to be a neurotoxin employed by the predators of the echinoderm *Diadema antillarum*. This is secreted by the salivary glands of the mollusc *Cassis* and is applied to the surface of the test by the manoeuvres of the molluscs' proboscis. As a result, the shadow response of the spines is abolished and the animal falls an easy prey.

Burn Toxin

It is most unlikely that in nature an echinoderm, if it is exposed by the tide, would ever reach a very high temperature. Consequently, the final toxin to be considered is of a more theoretical interest and probably does not play a part in the day-to-day existence of the animal. In 1958 Chaet and Cohen[119] made the interesting discovery that the peritoneal endothelium of the starfish *A. forbesi* would, when heated to 76°C, yield an extract which upon injection into a recipient animal caused autotomy of one or more of its rays and finally death. Animals scalded at 76°C for 1·5 to 2 min and then left, will autotomize.[110,126] Injection of the coelomic fluid of such animals into recipients also caused autotomy as the endothelial extracts had done. Heating the coelomic fluid alone showed that neither it nor the coelomic cells were the source of this toxin. Some of the experimental animals failed to autotomize and even survived a second injection of fluid. Incubation of a mixture of "resistant" and "active" fluids failed to demonstrate the presence of any anti-toxic factor in the resistant animals. Other tissues have also been shown to produce this toxin after scalding, the most effective being the haepatic caeca. Activity of these extracts was lost after dialysis against sea water, which suggests that the active principle is a rather small molecule. This reaction bears distinct similarities to the burn shock reaction common in mammals, where injection of extracts of burnt skin have been shown to lead to the death of the recipient animal. The onset of shock in cases of burning in man is possibly associated with changes in the histamine content of the blood.[677] The exact mechanism for the release of the histamine is not clear, but it almost certainly results from some sort of surface reaction by the burn toxin. The presence of histamine in the tissues of echinoderms has been recorded,[507,508] and a pharmacodynamic action has been shown to occur when histamine is applied to the longitudinal muscle of holothurians.[222] Evidence has also been presented to suggest that the release of histamine follows treatment with holothurin. Some of the physiological pre-requisites for a typical shock reaction are therefore to be found in echinoderms, but its function in this particular context is not clear.

Immune Response

The anaphylactic shock reaction following upon injection of an antigen into a previously sensitized animal almost certainly follows a similar pattern to that just described. It is of interest, therefore, to find that echinoderms exhibit the elements of an immune response. Extracts of amoebocytes of *A. forbesi* will produce clumping when injected into the coelom of another animal,[38] and antibodies against the animal's own tissue were found in *A. rubens* with regenerating rays.[799] No such antibodies were to be found in normal healthy starfish. It seems strange that the process of regeneration is accompanied by the appearance of autoantibodies, and their function is not understood. The perivisceral fluid of many invertebrates contains substances which react with their own tissue extracts, but true precipitation reactions cannot be obtained from echinoderms unless the cells of the perivisceral fluid participate; hence it can be best regarded as an agglutination.[418] The presence of at least four heteroagglutinins in the body fluid of *Patiria miniata* has been claimed by Tyler.[773] Phagocytosis and encapsulation appear to play a minor role, following the implantation of foreign tissue into the coelom of *A. forbesi*. The dermal branchiae possess longitudinal and circular muscle and it would seem that certain branchiae become filled with foreign tissue. These branchiae are then constricted by the musculature and autotomized. Another loss of foreign implants was thought to occur through the walls of the cardiac stomach. No distinction could be made between autologous and homologous tissues nor even from closely related members of the same family, but between members of different families destruction of the implants occurred. It would thus appear that coelomic cells possess an immunological competency of a kind to be found in the mammalian lymphocyte. This is a further example of the close physiological parallel between echinoderm amoebocytes and the various vertebrate cells. As has been noted in Chapter 2, the perivisceral fluid is normally a sterile medium as a result of the phagocytic activity of the cells.[38] In the light of such current work on the immunology of echinoderms, the possibility that this sterile condition could result, at least in part, from an antigen–antibody reaction cannot be ruled out. It is pointed out that the clotting reaction of asteroids and holothurians is weaker than in echinoids.[82] A reason for this may be found in the fact that both the former groups can effect a muscular closure of an injury to prevent loss of coelomic fluid, whereas an echinoid must rely upon its clotting reaction for this purpose. Possibly immune response within this phylum may follow similar lines. Echinoids appear to have evolved an immunity to their own toxins and such a response may be of use to combat bacterial injection following upon an injury. This type of reaction would also benefit those species which undergo autotomy.

Millott[525] has recently emphasized the part played by the axial organ in the immune response of echinoids. Peripheral cavities exist in this structure which are in communication with the perivisceral coelom. These cavities often contain amoebocytes aggregated into cysts which rotate. Injection of coelomic fluid of *A. forbesi* into *Strongylocentrotus droebachiensis* resulted in numerous cysts appearing which seemed to be clumps of *Asterias* coelomocytes. This was confirmed using more readily identifiable cells such as *Arbacia* sperm and a culture of the ciliate *Uronema*. It seems clear that the axial organ can in some way take up cells foreign to it, but their fate is unknown. Cysts which also rotate have been observed in the coelomic fluid of *P. miniata*.[376] These cysts contain phagocytic elements together with propellant flagella, which these authors suggest arise from the flagellated cells of tissues adjacent to the coelom. In hanging drop culture, these cysts remain viable for up to 74 days and their origin is obscure.

CHAPTER 7

SENSORY PHYSIOLOGY

Part 1. Gravity, Righting Reaction and Response to Mechanical Stimuli

THE echinoderms are surprisingly lacking in well-differentiated sense organs. Many of the animals' responses are due to the operation of independent effector organs and no clear evidence is available to suggest that echinoderms possess any significant learning ability. Simple conditioned reflexes, however, are claimed to have been stabilized[727] and are further considered in Chapter 11. Complex behaviour patterns, therefore, are absent and the animal reacts in a simple and often direct way to a wide variety of stimuli such as gravity, mechanical stimuli, and so on.

Crinoidea

Generally speaking, it seems true to say that echinoderms respond to gravity and endeavour to maintain a constant position in relation to it. Statocysts formed upon the classical pattern have been observed only in the synaptid, elasipod and molpadonian holothurians, although a somewhat abberant form, called sphaeridia, is to be seen in some echinoids. Gravity may not be without some effect in the production by the sessile crinoids of their typically erect form, but unfortunately the factors affecting these animals' growth are unknown. The position assumed by the free living Comatulidae seems often to bear no relation to gravity.[140,141] However, these animals creep or swim with the oral surface uppermost and when placed in an inverted position can perform the familiar righting reaction, although no specific gravity receptors have been described. The basis of this reaction in crinoids was investigated as long ago as 1884 by Marshall.[497] He showed that whilst the aboral surface of the calyx was insensitive to both mechanical and chemical stimuli, stimulation of the other surfaces led eventually to flexion of the arms over the disc. Stronger stimuli resulted in more pronounced movements and finally to detachment of the cirri and flight of the animal. It was further demonstrated that evisceration had little effect upon these responses or upon the righting reaction, thus ruling out the possibility that gravity might be perceived through its exerting a pull upon the viscera. The main pathways of communication must therefore be the pentagonal commissure, in which conduction is strongly decremental, and through the decussating fibres leading to the brachial nerve. Destruction of the aboral nerve centre leaves the animal in a completely motionless state, although individual arms will respond to touch. After destruction of the soft tissue, conduction in these arms is still possible in a proximal direction, due to the existence of the brachial nerve embedded in the brachial plates. *Antedon* has considerable acrobatic activity and in view of the foregoing it seems most likely that positive stereotropism of the cirri is responsible for initiating the righting reaction. The chief feature of this process is the lifting clear of the substratum of the disc, followed by the recurving of the arms to secure an anchorage, the animal subsequently pulling itself over. Contact stimuli inhibit the righting reaction and contact with the oral surface inhibits the grasping reflex of the cirri.[548] The opposing flexure of each arm of the arm pairs is suggestive of some form of central inhibition and this was found to be abolished by dilute solutions of strychnine.[545] The effect of a strong stimulus upon a fatigued animal was similar to that of a weak stimulus upon a fresh active one. The aboral nerve centre is mostly concerned with the initiation of the appropriate responses upon receipt of a given stimulus,[447] and conduction can be via a variety of pathways.

Holothuroidea and Statocysts

In the synaptid, elasipod and molpadonian holothurians, statocysts have been located in varying numbers around the point of emergence of the radial nerve from the calcareous circumoesophageal ring and the organ receives a branch from this nerve. Basically these statocysts consist of a vesicle made up of flattened non-ciliated epithelium enclosing up to 20 lithocytes which are vacuolated cells containing the lithospheres themselves. It is undoubtedly these structures which are responsible for the mediation of the very clear-cut positive geotropism found in many of the burrowing holothurians.[604] Statocysts and tactile sensitivity generally may be responsible for maintaining an equilibrium in the synaptid *Opheodesoma spectabilis*, but seemingly there was no gravity orientated behaviour.[52] Isolated tail portions of *Caudina chilensis* are, however, negatively geotropic as also are several climbing species.[815,816] Some forms without statocysts, such as *Cucumaria cucumis*, are also positively geotropic, but the mechanism for the perception of gravity is unknown. The application of centrifugal force results in the anterior end taking up a centrifugal position which is what would be expected in a burrower.[817] Exposure may also play a part in the burrowing of some holothurians and the thigmotactic response in general will be of importance in this respect and also with regard to the righting reaction, a reaction which is also carried out by the apodous forms. When buried, the animal takes on a flaccid condition, but upon exposure the body wall contracts and the animal takes on a more rigid form.[816]

In the epithelium of holothurians are numerous elongated cells whose processes connect directly with the nerve net. These neurosensory cells are probably sensitive to touch and chemical substances and Olmsted[604] called them a "universal sense organ". If these neurosensory cells can only signal the degree of stimulation into the nervous system by the normal method of modulating the frequency of impulse discharge, then obviously this will not permit the animal to discriminate between the various modes of stimulation. Within limits, this situation obtains in the protozoa and there is no *a priori* reason why it should not exist in a single cell of a multicellular animal. Of necessity, therefore, the behavioural responses will be simple and stereotyped. This is certainly the case in the holothurians where both mechanical and chemical stimuli to the trunk result in an inward dimpling of the body wall. Upon more intense stimulation the animal releases its hold and moves away. The tail end and the tentacles are the most sensitive, the latter being withdrawn upon prolonged or intense stimulation.

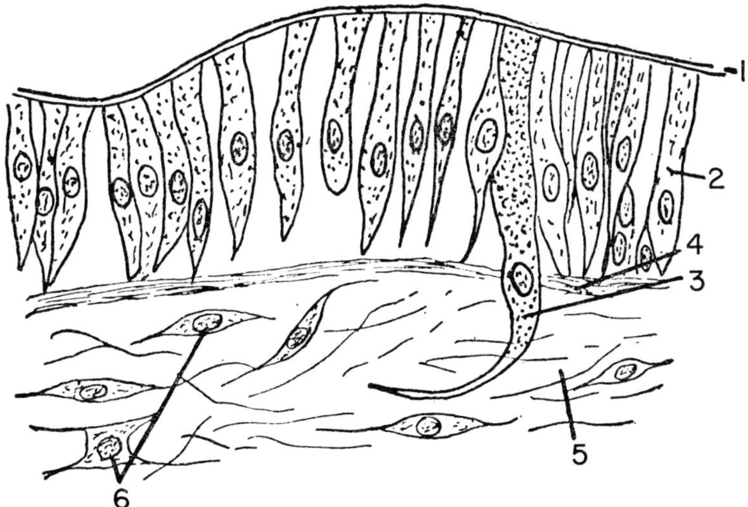

FIG. 10. Section through the body wall, near the anus, of a synaptid showing the sub-epidermal plexus. 1, cuticle; 2, ordinary epidermal cells; 3, gland cell; 4, sub-epidermal plexus; 5, dermis; 6, connective tissue cells.

However, they appear to be less sensitive when the animal is moving.[168] Thigmotaxis as a means of orientation is more important in some species than in others. Generally speaking, sea cucumbers are positively thigmotactic and this leads to many well-known behavioural characteristics: such as attachment of the tube feet, withdrawal of the animal into crevices and assists in or initiates the righting reaction. An analysis of the particular contribution made by each of these responses in *Thyone* was made many years ago.[616] Some sea cucumbers respond to mechanical stimuli by hardening of the body wall, and this is accompanied by a shortening of the body and appears to be mediated by a nervous reflex which is curiously insensitive to both electrical and chemical stimuli.[709] Such a condition is not unknown amongst other echinoderms. As the integument is composed solely of collagenous fibres, hardening must be due to a sliding of filaments rather than an actual shortening. The function of the glandulosensory warts which are found in the apodous synaptids is not known with any certainty. They possess sensory cells with fibres running to a ganglion in the epidermal nerve net.

Echinoidea

The spheridia of the Echinoidea (except Cidaroidea) are spherical transparent bodies situated either in or near the ambulacrum in varying numbers. Many echinoids are negatively geotropic and will ascend the walls of an aquarium. Ablation of these spheridia in *Psammechinus miliaris* does not destroy the response to gravity, but it does upset the general muscular tone of the animal.[67] Experiments upon the effects of extirpation of these structures indicated a loss of podial co-ordination and hence a prolongation of the righting reaction.[192] Gravity is probably perceived by some other mechanism, for example tension in the mesenteries, and the spheridia exert a trimming function. The burrowing spatangoids and clypeastroids are probably positively geotropic, but experimental proof seems wanting, although when placed upon sand they immediately commence to burrow. The righting reaction in regular echinoids is carried out by the attachment of the appropriate podia, which then contract and eventually the test is pulled over. Even portions of the test have been found to give this reaction (Romanes and Ewart[676]). In spatangoids such as *Echinocardium*, the righting reaction is quite rapid and most use is made of the spines. These animals are able to right themselves even when buried. Keyhole scutellids such as *Mellita quinquesperforata* are said not to right themselves but to rely upon wave action to re-orientate them.[394] It is possible that the lunules are essential for this purpose, for in *Astriclypeus* blocking them with soft paraffin wax resulted in a threefold increase in the burrowing time and rendered the animal incapable of righting.[363]

The tube feet, spines, pedicellariae and general body surface are sensitive to touch, there being present in the epidermis the familiar neurosensory cells.[317] These also occur in localized patches inside the jaws of the globiferous pedicellariae. A weak mechanical stimulus results in the localized pointing of spines towards the stimulated area, this being the main defensive reaction of the animal.[774,775,777] More intense stimulation of the spines cause them either to become rigid upon the test or to point away from the source of stimulation, thus exposing the pedicellariae for action. These pedicellariae are controlled by an interesting muscular system and their mechanism is dealt with in Chapter 11. Intense stimulation results, as always, in flight of the

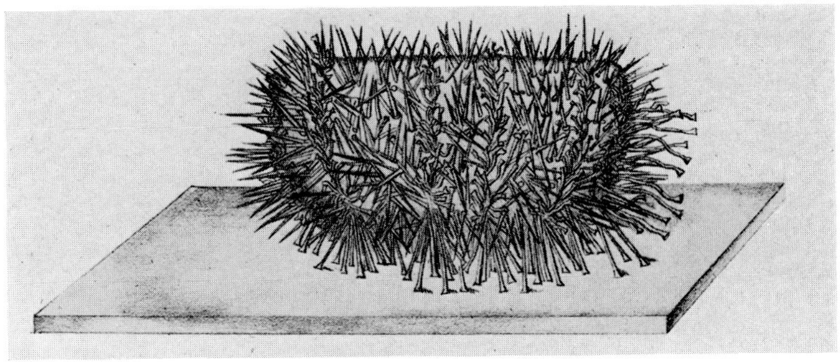

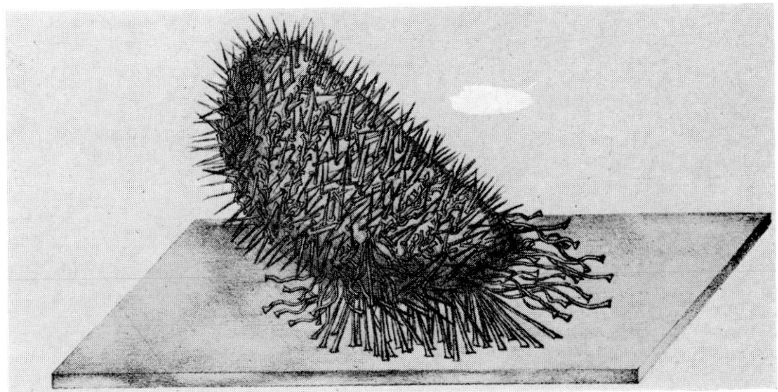

FIG. 11. (i) Righting reaction of *Echinus*.

(*facing page 84*)

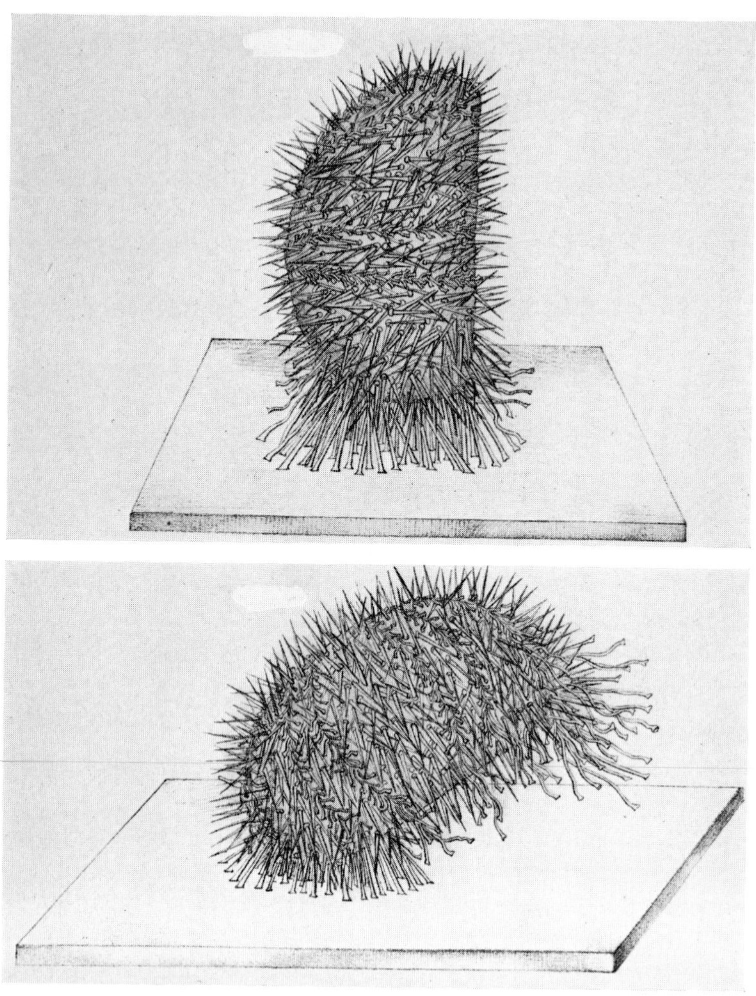

FIG. 11. (ii) Righting reaction of *Echinus*.

animal. When sea urchins are moving, many tube feet are protracted at the leading edge to act as feelers. A slight mechanical stimulus will cause them to retract and a stronger stimulus will in addition cause the protraction of feet of another radius with the result that the animal moves off in another direction. Jarring will cause many podia to adhere firmly, with the result that the animal can withstand a degree of wave action.

Asteroidea

Once again, the Asteroidea possess numerous neurosensory cells, between 4000 and 70,000 per square millimeter having been estimated;[717] they probably fulfil both tango- and chemosensory functions. Asteroids lack any specialized gravity receptors and they do not appear to respond to gravity *per se*. However, *Asterina gibbosa* can be shown to be negatively geotactic in that it always ascends a water-filled glass cylinder from which all air was excluded.[172,173] As with echinoids, this response was given by portions of the animal and it was suggested that the weight of the animal stretching the podia is the basis for this reaction. In an attempt to artificially lighten the body by attaching corks to it, the resulting tension developing within the tube feet caused the animal to behave in a positively geotactic manner. The familiar righting reaction is given by all asteroids, and because of the variety of shapes to be found within this group it is not surprising to find that the details vary considerably from species to species. This reaction has received the attention of a number of workers,[199,267,378,487,543,544,546,547,597,640,676,684,812] and is well reviewed by Hyman.[362]

Notwithstanding this variety of forms which the righting reaction can take, three basic movements emerge.[597] When placed upon the aboral surface, the arms of an asteroid will, after a short quiescent period, start to curl aborally. This is the result of the animal executing the dorsal reflex.[485,487,546] This results in the aboral curling of the arms when an animal in the normal position is stimulated upon its aboral surface. When the tips of two or three arms have come into contact with the substratum, the tube feet commence stepping, bringing these particular arms underneath the disc. The tube feet of the remaining arms detach themselves if necessary and the arms are brought slowly over the attached portion of the animal. The entire oral surface of the animal is in contact with the substratum. Folding over is only a variant of this somersaulting process. In the formation of a "tulip", the tips of all the arms are brought together underneath the animal. The body of the animal thus becomes raised up and eventually topples over, whereupon the arms unfold and regain their correct position. The time taken for this procedure ranges from a minute or even less up to an hour or so. The pentagonal starfish are the less speedy at righting, but nevertheless are quite adept in view of their shape. *Pteraster tesselatus* has a supradorsal membrane and the cavity so formed between this and the aboral surface of the body can be filled with water in various regions of the body so causing the animal to tilt, thus permitting the podia to gain a purchase.[674] The behaviour of *Culcita* is similar.[597] What is apparently a physiological bilateral symmetry exists in some starfish and, assuming the absence of any other directional stimuli, these animals always tend to right using the same arms. Such has been found in *Astrometis sertulifera*[371] and in *Pycnopodia helianthoides*.[406,407]

It will be realized that the righting procedure outlined above is extremely complex and demands a reasonably high level of co-ordination. The integrity of the circumoral nerve ring is essential for such co-ordination and in those animals where this has been transected the righting reaction is a series of unco-ordinated twisting movements of the arms, although the animal still ends up with its oral surface in contact with the substratum. The nature of the stimulus which evokes the righting reaction has been and still is a matter of debate. Some authors suggest a central nervous origin for such a stimulus,[371,640] whilst others incline to the view that it is peripheral. The part played by the positively stereotropic behaviour of the tube feet may also be important.[543,546,684] A really securely attached arm can inhibit the righting reaction of the others.[472] In addition to the contact stimulus with the substratum, the pull of the viscera within the coelom has been suggested[812] and denied[199,267] as the releasing stimulus. Positive stereotropism is an important factor in other echinoderm classes in maintaining the animal in its correct posture and is doubtless important here too.

The surface of asteroids is covered with a variety of appendages including spines, pedicellariae, thin-walled projections such as the dermal branchiae or papulae and the tube feet. Of these only the tube feet have been intensively investigated and these are considered in Chapter 8. A classical overall account of the three former types of appendage has been provided by Jennings,[371] which has been added to in later years. He concluded that the pedicellariae were mainly concerned with the removal of detritus and small animals from the integument, the latter being eventually passed to the mouth. To facilitate this behaviour the spines bearing the pedicellariae will often bend towards the source of stimulation and additionally rosettes of pedicellariae can be erected. Once closed upon an object they can retain their grip for considerable periods of time and are refractory to any further stimuli. In *Astropecten*, which lacks pedicellariae, pairs of opposable spines carry out a similar function to the jaws of the pedicellariae.[554] These localized responses of the spines and pedicellariae are mediated through the nerve plexus of the dorsal integument.[719,723] This plexus exhibits to a remarkably similar degree the diffuse and decremental conduction which Pantin[612] demonstrated for coelenterates (see Chapter 11).

The dermal branchiae are remarkably sensitive to mechanical stimulation and contract readily in response to a variety of disturbances. Such responses are eliminated if a cut is made parallel to the major axis of the arm, thus disturbing the transverse through conduction pathways in the nerve plexus described by Smith.[719] The function of the papulae would seem to be mainly respiratory and as a pathway for the loss of amoebocytes laden with excretory matter.

Experiments designed to investigate the gross responses of asteroids to mechanical stimulation have been the subject of a number of studies.[378,484,640,723] Mechanical stimulation of the podia causes their retraction and should such stimulation become more intense, the reaction spreads throughout the arms and the animal eventually moves off. Light mechanical stimulation of the remaining surface of the arm results in adjacent podia bending towards the point of stimulus. The neural pathways of this response and the physiology of the podia themselves will be considered in Chapters 11 and 8 respectively.

Ophiuroidea

The range of surface appendages in the Ophiuroidea is limited by comparison with the two groups already considered. Pedicellariae and papulae are entirely wanting and the epidermis is restricted to small areas at the ends of the arms, or in some species it too may be completely lacking. However, a light touch on the side of the arm will result in that arm taking up the escape position, namely, with the base curved away from the stimulus and the tip of the arm recurved towards the stimulus. The tip of an arm will often sieze an object in its path, such a movement being the one typically employed in the snake-like process of locomotion.[486,502] Burrowing is less common in ophiuroids and their reaction to gravity seems to be limited to keeping the oral surface in contact with the substratum. The burrowing of *Amphiura chiajei* seems to be more the result of a negative phototaxis than a positive geotaxis, the podia being the main organ responsible for the movement of sand.

In addition to the stereotropism of the tube feet, the body also is strongly stereotactic and this serves not only to assist in a more speedy righting reaction than is common in most echinoderms, but also to account for their predilection for crevices and corners.[163] The dorsal reflex described for asteroids is also given by ophiuroids. When roughly handled these animals assume a characteristic rigor or "freeze" attitude, which is also a feature of certain echinoids. The righting reaction, which may be accomplished in as little as 2 sec, is executed by the spreading out of two arms at 180° and then using the other three to push the animal over. Variations of this procedure have been described by several authors, some of which approximate to the asteroid folding over. Although no bilateral symmetry is to be observed in this reaction, the exact direction in which the animal will right often depends upon the incidence of other stimuli. Portions of the animal will also perform a modified righting reaction.

As with asteroids, the nature of the triggering stimulus in the righting reaction is a matter of some debate. Wolf,[812] using *Ophioderma*, contended that tension in the visceral mesenteries was responsible and showed that by filling the stomach with iron filings and applying a magnetic force in the aboral direction a normally orientated animal could be persuaded to invert. Such an experiment, of course, does not limit the tension to these mesenteries, for being a fairly rigid animal, at least some will be transmitted to the podia. Removal of the stomach did not impair its righting ability[267] and, if supported in an inverted position by a small cone of sand under the disc, righting still occurred.[487] Animals falling through a column of water will right whilst falling, but will not do so if they are released in a state of rigor. Positive stereotropism would again seem to be the most likely explanation, although a conclusive experiment has yet to be performed. Whilst the spines are the most obvious appendage concerned with locomotion, being movable and mounted upon small tubercles in a similar manner to echinoid spines, the tube feet also play a hitherto unrecognized role[813] which will be further discussed in Chapter 8.

SENSORY PHYSIOLOGY

Part 2. Chemical, Thermal and Photic Stimuli

Crinoidea

Few observations have been made upon the effects of various chemical substances upon crinoids. The tropical crinoid *Tropiometra carinata* reacts to formalin, alcohol and magnesium sulphate and shows a differential susceptibility to carbon dioxide.[141]

Holothuroidea

A number of elaborate experiments have, however, been devised using holothurians, and the entire body surface of *Holothuria surinamensis* is sensitive to a range of salts, acids, alkalis, alkaloids and a whole host of organic substances including narcotics.[168] At the time it was thought significant, though not in any phylogenetic way, that these substances fell into the same sort of categories as do human taste sensations. The effect on the animal of all these substances was similar to that for mechanical stimulation, namely, retraction away from the point of stimulation. The differential sensitivity of the various regions of the body was parallel to that for mechanical stimulation which would tend to support Olmsted's idea of a "universal sense organ".[604] Both he and Yamanouchi,[816] who worked with *Caudina chilensis*, found that stimulation by cations and anions followed the familiar lyotropic series in the intensity of their effects: Cl > Br > I, and K > Na > Li. As would have been expected from our knowledge of the penetration of membranes by ions and molecules, they found that weak acids were more effective than strong ones at a similar pH. Differences in osmotic pressure, however, passed unnoticed in *H. surinamensis* but were readily perceived by *Synaptula hydriformis*.

Interesting as these results are in helping to delimit the range of sensitivity of the animal concerned and its sense organs, they give little insight into the role of the chemoreceptive sense in the normal behaviour of these echinoderms. Using organic substances and animal extracts it was found that synaptids would react negatively to a piece of decaying starfish, but no response could be elicited from *Thyone* using crab extracts or even mud from its usual habitat.[139,616] It is relevant to mention here that the synaptids, in common with other Apoda, possess both particularly well-developed glandulosensory warts over the general body surface, which are supplied with nerve fibres, and numerous pits along the tentacles. It is believed that these pits function as chemoreceptors and it is interesting to note that in common with other olfactory or chemoreceptive epithelia, they possess cilia or some form of cytoplasmic process.[179] Electron microscope studies indicate that basal regions of the epithelial cells of *Diadema setosa* are extended into threads which intermingle very closely with processes of the nerve plexus.[389] A similar circumstance would seem to exist in the gut of *Stichopus japonicus* where the peritoneal epithelium is similarly modified. Such work gives an anatomical basis for the "universal sense organ" and, incidentally, to Millott's description of the surface of *D. antillarum* as a "giant retina".[523]

Echinoidea

In contrast with the holothurians, very little is known about the chemosensory powers of echinoids and this has not been supplemented significantly since the original work of Romanes and Ewart[676] and von Uexkull.[774,775,777] The result of chemical stimulation is broadly the same as for mechanical stimulation. The spines react by pointing towards a weak stimulus and away from a stronger one, followed eventually by the flight of the animal. Intense activity of the spines is to be seen following immersion in sea water containing obnoxious substances, caffeine being particularly effective in this way. Similar results have also incidentally been obtained with asteroids. Weak stimulation of the aboral surface such as the application of a dilute acid gives rise to localized podial protraction in the same way as does mechanical stimulation. Direct application to the tube feet themselves of such substances results in retraction and stronger stimulation of the arm results in the animal moving away.[378,484]

Asteroidea

The nutritional level of the animal can modify its behavioural response to certain stimuli, notably food. *Asterias rubens* exhibited little reaction to food when freshly caught and presumably well fed. When hungry, however, it could be led around an aquarium with food held in forceps.[675] The observations of many other workers strongly support the view that asteroids are able to detect the presence of food chemically.[511,565,716] However, this response could not be confirmed in all the species examined.[284,378,640] When the starfish *Luidia* contacts a brittle star it performs a characteristic leaping movement as if in an attempt to smother it.[239] If the tips of the arms of a buried *Luidia* are touched by an ophiuroid it will quickly surface and follow it. As extracts of *Ophiura* elicit these responses up to a range of 10 cm it is obviously chemosensory in origin. Animals have also been known to engulf cotton wool soaked in animal juice. It is well known that large numbers of starfish will accumulate in lobster pots and the reason for their often relatively sudden descent is unknown. A good mussel spatfall will also sometimes result in a large number of starfish appearing in a district. Such conflicting evidence cannot allow one to decide whether the chemotactic response is important or not in the Asteroidea. The migration of starfish will be referred to again in Chapter 11.

No published observations appear to be available indicating whether asteroids can distinguish between waters of differing salinity, although they are reported to be able to differentiate between solutions of potassium, calcium and magnesium chloride.[631] The present author has found that pipetting fresh water onto portions of the aboral integument of *A. rubens* causes the animal to move away from the direction of application. Such movement does not take place when sea water is employed, ruling out the possibility of rheotaxis. The area of integument between the arms seems most effective in eliciting this response.

Ophiuroids

In experiments with pieces of crab or fish, the chemotactic responses of ophiuroids can be quite dramatic, often waving their arms and making directly for the food. The distance over which they can detect their food varies with its nature, but it is certainly up to a foot in *Ophioderma*.[197,565,640,778] Upon reaching the food, it is seized by the tips of the arms and it is at this stage that inert materials will be rejected unless camouflaged with animal juices. The animal will rapidly withdraw from obnoxious chemicals[312] and will exhibit pronounced reactions to changes in osmotic pressure, although there is no evidence that this results in any directed movement.[502] An almost complete absence of information makes it impossible to pinpoint the receptor involved, but the tube feet have often been suggested for they possess the most highly developed epidermis, the typical neurosensory cells of other echinoderms being absent from the general body surface.

Comparing the responses of echinoderms to mechanical and chemical stimulation, it will be seen that there is a close parallel between the two, both in the form and intensity of the response. It is often very difficult to measure the intensity of stimuli of this nature, and evaluation of the results must of necessity be somewhat subjective, especially in relation to terms like "weak" and "strong". Allowing for such limitations, a fairly well-defined change in behaviour occurs when the stimulus intensity is raised. Particularly noticeable in holothurians, echinoids and asteroids where the neurosensory cells are so numerous, weak stimuli usually cause movement of the tentacles, spines, pedicellariae or tube feet towards the source of stimulation. More intense stimulation results in a reversal of these movements and in the eventual flight of the animal. Superficially these appendages have the characteristics of independent effector organs, but with an underlying and more central form of control which becomes more apparent as the stimulus intensity increases. A recent discussion of such aspects will be found in papers by Bullock[99] and Ewer.[225] A vast amount of work upon echinoderm behaviour was carried out between 1890 and 1930 and whilst the facts which emerge are not in dispute, the rigorousness of the experimental regime and the interpretation could well bear reinvestigation. During this period ideas upon animal behaviour were strongly influenced by the Loeb school whose attempts to reduce all behaviour to a series of autonomous and unco-ordinated tactic and kinetic responses are now viewed less favourably. von Uexkull's[774] reference to a holothurian being a "republic of reflexes" is just such an example. Although relatively simple animals, echinoderms do show the beginning of the more complex form of associative behaviour called learning and this aspect will be dealt with in Chapter 11.

TEMPERATURE

Crinoidea

The reactions of echinoderms towards temperature gradients is virtually unknown. The comatulids from the Torres Straits to the north of Australia were shown to select warmer waters when given the choice, their optimum being around 26°C.[140] For the remainder of the phylum the results may be more properly described as thermal tolerance. A wide tolerance of some 15°C was found in the crinoid *Tropiometra carinata* and this species would recover from short exposures of up to 34°C.[141] The optimum for *Antedon petasus* from Norwegian waters was 11°C; temperatures above 14°C were often fatal.[303] This illustrates the close correspondence between the temperature optimum as found empirically and the normal environmental range.

Holothuroidea

In holothurians the responses to localized application of hot and cold sea water seem to consist mainly of relaxation or contraction respectively of the body wall musculature. The temperature tolerance of this group is wide, ranging from about freezing point to around 37°C.[168,604,616,816]

Echinoidea

The purple sea urchin *Strongylocentrotus purpuratus* tolerates sea water in the range 5°C to 23·5°C. Some acclimatization is to be observed at the lower end of the scale, the animals tolerating 1·9°C after 15 hr exposure and even righting at this temperature, but no such acclimatization was detectable at the upper end. The increase in oxygen consumption at the higher temperatures was, however, less than would have been expected. This falling off in the effects of rising temperatures may be due to the adaptation of the existing enzymes or their modification to lower activation energies, the reactive level therefore being reached earlier.[230] At the extremes of its range, the animal is sub-tidal, where it may be expected to avoid the greater fluctuations. The cold water populations of this species have a shorter gonadial growth period than those in warmer water.[74,77] A correlation between the heat death temperature and the period of exposure to various temperatures in *Arbacia* has been observed. This species lived for several hours at 35°C but died in 9 min at 42°C.[607]

Asteroidea

There is good reason to believe that the distribution of certain asteroids is in part controlled by temperature. On the eastern shores of North America *Asterias forbesi* extends from Maine as far south as Mexico and can withstand temperatures as high as 33·5°C, but at 35°C this species is rapidly killed.[607] *A. vulgaris*, on the other hand, is killed by temperature in excess of about 25°C and this species is restricted to the more northerly waters. In this respect it is more like the European population of *A. rubens*, for animals taken from the North Sea do not survive temperatures much in excess of 25°C either. The bipinnaria of the Japanese starfish *A. amurensis* would tolerate temperatures from 5° to 20°C, the young adults being a little more hardy, surviving to 26°C.[685]

Using isolated tissues, the thermal tolerance of the ciliated epithelial cells from 11 species of starfish from the Japan, Okhotsk and Barents Seas has been investigated.[711] The results confirmed what had already been found using whole animals, namely that the tissues from the warmer water species were more resistant to high temperatures than were those from a colder environment. When the water in which *A. rubens* was living was raised from 4°C to 10° or 11°C, destruction of the stomach tissue occurred in some animals. Their reproductive behaviour was apparently also interfered with.[728]

Ophiuroidea

It is unfortunate that with the exception of Orr[607] no comparable experiments upon the behaviour and tolerance to changes in temperature in ophiuroids have been carried out. He found that *Ophioderma* died in 27 min at 37°C and in 9 min at 42°C. The highest temperature at which the animal could survive is not stated. Some species have an almost universal distribution, such as *Amphipholis squamata*, with an apparently uniform morphology. Either these species have an enormous temperature tolerance or they have split into a number of physiological races with differing thermal tolerances.

PIGMENTATION

The echinoderms are an extremely colourful group, animals of almost every hue being recorded. The pigments may be present in the epidermal layers of the body wall or may be deposited in a more permanent form in the skeleton. Although so often highly coloured, in this phylum the physiological colour changes are restricted to diurnal variations, usually of small magnitude compared with the rapid alterations of colour to be seen in the molluscs and crustaceans. Morphological colour change due to the deposition of pigmented excretory material with increasing age is found to occur notably in the crinoids and holothurians. The absence of any rapid colour change mechanism is doubtlessly correlated with the lack of any high-speed co-ordinating system within the phylum. Those pigments associated or thought to be associated with respiration are considered in Chapter 9.

Crinoidea

The littoral crinoids from tropical waters are acclaimed by many zoologists to be the most beautiful of all marine animals. Be this as it may, the variety of pigments definitely identified from this class is small. Older workers isolated a number of coloured substances, many with indicator properties, the chemical nature of which is only now becoming known. As early as 1877, Moseley described "red" and "purple" pentacrinin from the stalked crinoids and "antedonin" from a comatulid.[557] His work has now been supplemented by other workers.[2,425,569] Dimelow tentatively identified a hydroxynaphthoquinone from *Antedon bifida* and also recorded the presence of both free and esterified astaxanthin, xanthophyll and β-carotene.[200] The possibility of some of these pigments being of dietary origin cannot be excluded, for this crinoid is a detritus feeder and ingests small copepods. This species was found to react most strongly to green light and may also possess light-sensitive nerves. The work of Sutherland and his colleagues has done much to elucidate the chemical structure of crinoid pigments.[638,639,744,745] From *Comatula pectinata* a mixture of anthraquinoid pigments was obtained, the main component being named Rhodocomatulin which chemically is 4: butyryl-1.3.6.8-tetrahydroxyanthraquinone. A derivative of this substance, rubrocomatulin, was also characterized. From the species *Ptilometra australis* and *T. afra* a variety of anthraquinone pigments was obtained including rhodoptilometrin, ptilometric acid and isorhodoptilometrin, this latter substance being the red-orange pigment described by MacMunn[569] in 1889. It is rare for pigments of this type to occur in animals, the only other important record being from the Coccidae. An interesting instance of the stability of this class of pigments has been provided by Blumer.[63,64] He extracted crystalline organic pigments called fringelites from fossil crinoids of the Upper Jurassic period. These substances were found to occur mainly in the rooting systems where it seems likely that the more anoxic conditions which obtain in deep bottom ooze facilitated their petrification. From *Apocrinus*, fringelite D, E, F and H were obtained and characterized. Chemically they represent a reduction series of increasing stability with the consequent increase in their concentration in the fossil. Further reduction proceeds only very slowly, the products being hydrocarbons. The concentration of pigments is some 200 times that of hydrocarbons. Distribution of the pigments suggests that present-day stalked crinoids are closer to the fossil crinoids than are the comatulids. Other components include quinoids resembling aphin and hypericin found in certain arthropods. The presence of resonating quinoid structures within a pigment is often indicative of some photochemical activity and it comes somewhat of a surprise to learn that crinoids have little reaction to light, although it is reported that *T. carinata* is weakly phototactic.[141]

Holothuroidea

Many species of holothurian are dark in colour and there is very strongevidence that in *H. forskali* and *Thyone briareus* at least this is due to the presence of melanin.[513,515,516] Coelomic fluid from these species when left in contact with air forms a stringy clot which slowly turns brown and, after 12 to 24 hr, black. This is due to the formation within the coelomocytes of melanin. The reaction takes place on the surface of the large spherules which contain an enzyme system with the properties of a phenolase and possibly also the pigmentary precursor. *In vivo*, this pigment is transported and deposited in the body wall after breakdown of the cells. It has been suggested that such pigment is re-utilized by other amoebocytes,[155] although no recent evidence to support this contention is available.[516] Early descriptions of lipochromes (carotenoids) were made by MacMunn[569] and since his time numerous surveys have been carried out.[473–476,488–490] Carotenoids are reported from the body wall, much of the viscera, especially the gonads, the females generally having a higher concentration than the males. These pigments include xanthophyll, β-carotene and astaxanthin (astacene). The red holothurian *Stichopus tremulus* possesses no carotene, but the ketonic provitamin A (echinenone) which although found commonly in echinoids has only been described in one asteroid and incidentally a sponge.[207] Pigments with a green fluorescence have also been extracted from a species of *Holothuria*.[59,54,796]

Holothurians are frequently nocturnal animals and the tentacles of the dentrochirotes are fully expanded at night. Such behaviour is a reflection of the general negative phototaxis exhibited by the group. The reported greater degree of activity in wintertime may be a response to either photic or thermal conditions. *Thyone* reacts only to a sudden decrease in illumination,[616] a situation shared by *H. surinamensis*[168]

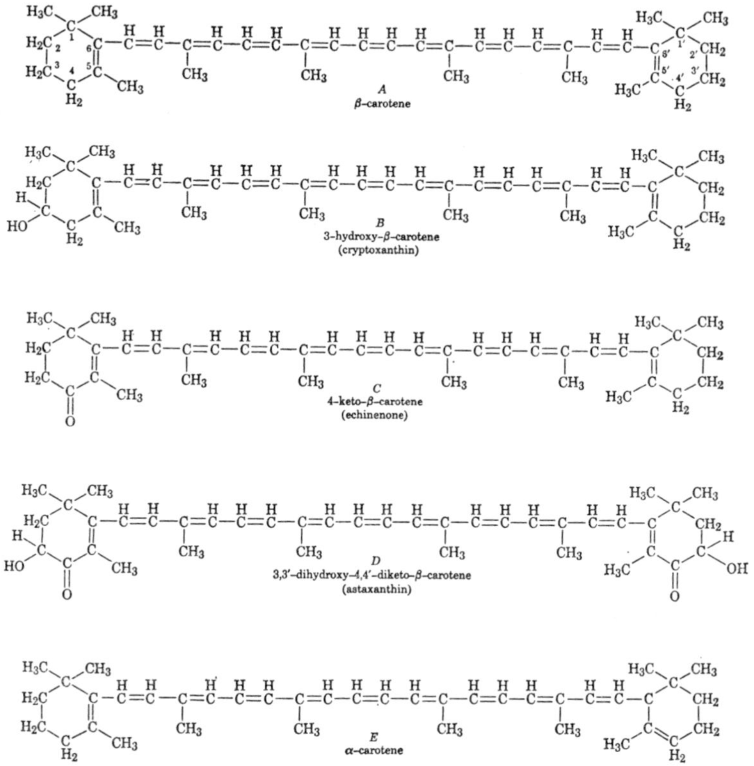

A
β-carotene

B
3-hydroxy-β-carotene
(cryptoxanthin)

C
4-keto-β-carotene
(echinenone)

D
3,3′-dihydroxy-4,4′-diketo-β-carotene
(astaxanthin)

E
α-carotene

FIG. 12. Some carotenoids occurring in echinoderms.

and recalls the more thoroughly investigated shadow response of certain echinoids. However, *Caudina chilensis*,[816] *H. poli*[344] and *Synaptula hydriformis*[604] react only to an increase in illumination. The whole surface of the body is light-sensitive, although paired structures akin to eyespots have been described, near the base of the tentacles in certain synaptids, which are supplied by a branch of the tentacular nerve. The response to light consists of tentacle retraction followed by a negative photokinesis. The period of illumination required to produce these effects (also given by isolated tentacles) may be extremely short. It would thus seem that the pigments of holothurians serve chiefly in a protective way rather than as a photochemical one, although in some sensory cells in the epidermis they may assist in the absorption of light. *Opheodesoma spectabilis* has a distinct diurnal rhythm dependent upon the changes in light intensity.[52] This animal is also photonegative and it is claimed that its ocelli are involved. Two colour groups exist in the species *Cucumaria curata*; the lighter ones, some of which may be nearly colourless, are the more sensitive to light.[715] No evidence is available to show whether holothurians undergo any morphological colour change as a result of prolonged exposure to light or increasing age.

Echinoidea

The pigments of the echinoids appear to fall into three categories—carotenoids, naphthoquinones and melanin, and additionally a lipofuscin may be present.[580] The presence of a range of carotenoids from the gonads of six species of echinoid, both regular and irregular, has been reported.[263,456,476] Isolation of α- and β-carotene by Lederer[456] led to the characterization of a carotenoid exhibiting vitamin A activity (see ref. 457 for review). The extensive work of Fox and his colleagues (see ref. 262 for review) has extended the work to include a variety of carotenoids, many of which are of plant origin—zeaxanthin, diatoxanthin, pentaxanthin, petuloxanthin (= sulcaxanthin = peridinin), antheroxanthin, lutein and several new xanthophylls, but little trace was to be found of esters. He points out that the females of *Strongylocentrotus purpuratus* have a higher carotene content than the males, whereas in *Dendraster excentricus* and *Lytechinus pictus* the converse is true. In general it is the viscera which contains the majority of these pigments, principally in the gonads.

The echinoids are one of the groups of animals to contain rarer pigments of the naphthoquinone type and MacMunn[566] first gave the name echinochrome to one such pigment from *Echinus esculentus* without knowing its chemical composition. Although their occurrence in the crinoids has already been noted, it is in the echinoids that they reach their widest distribution and diversity. In contrast to the carotenoids, which seem to be found mainly in the viscera, naphthoquinones are present in both soft and skeletal components of the animal. The nomenclature of these pigments is in a rather confused state owing to the inadequate characterization of the earlier described pigments, hence the occurrence of frequent synonymies.[309,310] The first pigment to be accurately characterized was isolated from *Arbacia pustulosa* and called echinochrome A, which chemically is 2-ethyl-3.5.6.7.8-penta-hydroxy-1:4-naphthoquinone.[802]

This substance has since been recorded from the skeleton or tissues of a number of species including *A. aequituberculata* and *Paracentrotus lividus*,[305] *L. pictus*, *S. purpuratus* and *D. excentricus*,[263] *Diadema antillarum*[522] and *D. setosum*.[590] The related echinochromes B, C and D have also been described, all from the ovaries of *A. pustulosa*.[305,428] The relative abundance of these pigments varied during the year, echinochrome A being maximal in the spring whilst B and C were maximal in the autumn. The echinochromes are usually red or reddish-orange in colour and in the case of *D. antillarum* have a fairly wide-spread distribution within the animal. Significantly it occurs as droplets or granules throughout the tissues or within the amoebocytes as sphaeroids to which granules may be closely applied. This pigment, like the melanin of both holothurians[516] and echinoids,[370] may be deposited in the tissues by disintegration of the amoebocytes.[522] Such cellular disintegration occurred *in vitro* and it has been noted that when liberated some of the red pigment turned brown or black.[522] It is thus possible that the black pigments of echinoids may have a dual origin, from a naphthoquinone and from a melanin, the former possibly existing in a partially reduced state *in vivo*.

A group of closely related pigments which would seem to occur exclusively in the spines and test of certain echinoids have been termed the spinochromes. The original formula[428] has been confirmed recently.[310] Goodwin recognizes only two spinochromes, A and B, which are violet and bluish-green respectively and were demonstrated to be present in *E. esculentus* and *P. lividus*.[310] These substances are able to form coloured lakes with mineral salts such as calcium within the skeleton of echinoids. Colour variations to be observed within these species, and ascribed to variations in the proportions of spinochrome and echinochrome,[305] are in fact due to this cause, for the presence of echinochrome has not been confirmed in the test or spines of either animal.[310]

Pigment	Structure	Synonyms
Spinochrome A	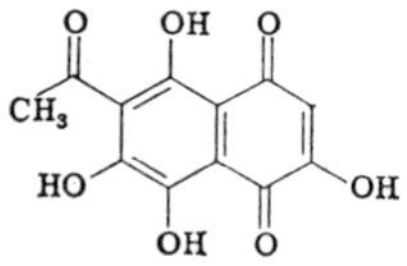	Spinochromes M, Mg, Aka_2, P
Spinochrome B	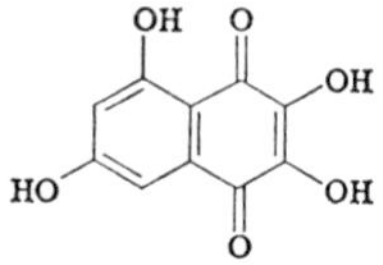	Spinochromes N, P_1, B, M_2
Spinochrome C	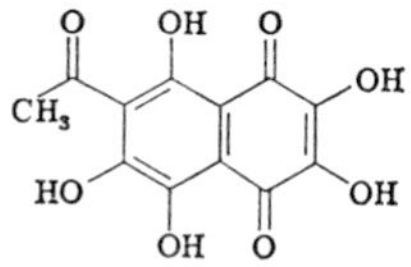	Spinone A, Spinochromes F, F_1, M_3
Spinochrome D	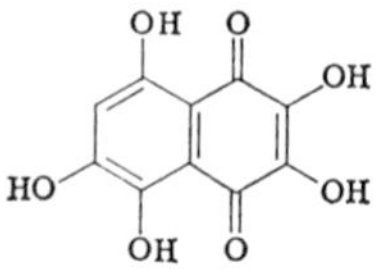	Spinochromes Aka, Aka_1
Spinochrome E	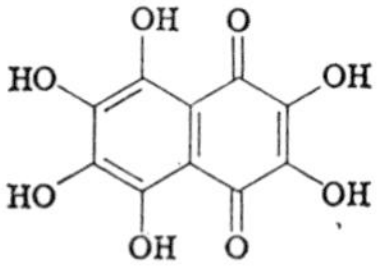	
Spinochrome G	?	? Spinochrome B

FIG. 13. Naturally occurring spinochromes in Echinoidea.

Various other spinochromes must now be recognized, although the possibility of synonymies must not be overlooked. In a long series of papers Kuroda and her colleagues[430–440] have described a number of spinochromes from various sea urchins. Separate pigments which they extracted from the test and named testachromes proved to be identical with certain spinochromes. Spinochrome Aka_1 and Aka_2 were isolated from *Pseudocentrotus depressus*; B_1, B_2 and possibly an additional B_3 from *S. pulcherrimus*; M_1 and M_2 from *Anthocidaris* (=*Heliocidaris*) *crassispina*; and F_1 and F_2 from *Heterocentrotus mammilatus*. B_1 and M_2 are identical compounds and so too are M_1, Aka_2 and B_2. Goodwin and his colleagues equate spinochrome Aka with spinochrome D. Lederer[458] shows that the violet to green colour of the North Atlantic *Paracentrotus lividus* is due to the presence of spinochromes A and B, but that the brownish tints of the Mediterranean animals were due to the presence of spinochromes B, C, E and G. He also confirms the existence of spinochrome P in this species, first described from *Arbacia pustulosa*.[305,559] Spinochrome E has also been shown to exist in *Psammechinus miliaris*[834] and spinochrome M is also said to occur in *Anthocidaris crassispina*, *Hemicentrotus pulcherrimus* and *Echinarachnius mirabilis*.[822] A further spinochrome, spinochrome H, has also been described recently.[128] Some recent work has indicated the existence of larval naphthoquinones quite distinct from those in adult animals in *S. droebachiensis*.[315] It is thought that the adult later develops the necessary enzymes to synthesize the adult pigments. Hybrid larvae of *droebachiensis/purpuratus* also possess small amounts of echinochrome C and D, neither of which occurs in the adults of either of the parent species. A spinochrome from the spines and test of the echinoid *Salmacis sphaeroides* has been chemically identified as 3-acetyl-2.5.6.7-tetra hydroxy-1:4-naphthoquinone.[311]

Echinochromes were originally considered to be respiratory pigments functioning in much the same way as does haemoglobin.[567] This view was later modified and it was thought to behave more as an activator than a carrier.[106] A claim by Friedheim[272] that echinochrome stimulates both the unfertilized eggs of *Sphaerechinus granularis* and rabbit erythrocytes by a factor of about fifteen was later criticized on the grounds that a tissue brei rather than a pure preparation of echinochrome was used.[772] A claim for sperm-activating properties[327] has been similarly denied.[156,772] Fox[262] was of the opinion that echinoids would ingest, along with their plant food, much naphthoaromatic material, which after metabolic conversions of one sort or another may be stored temporarily prior to excretion. Much excretory material is eliminated by way of the integument and it is possible that once in this region of the animal it might be incorporated into the test or spines as the calcium salt, the naphthoquinones being weakly dibasic. If a dietary origin for these pigments is correct, and diet does seem to affect the colour range exhibited by some species such as *Dendraster*,[308] then these pigments should also occur in other herbivores possessing a calcareous skeleton, which does not appear to be the case. However, there is an interesting case of a carnivore feeding upon *Strongylocentrotus droebachiensis*.[262] The sea-otter *Enhydra lutris* has its bones coloured purple by the calcium salt of a polyhydroxy naphthoquinone with chemical similarities to the pigment "X" of Goodwin and Srisukh[310] which they extracted from *Echinus esculentus* and *Paracentrotus lividus*. Chemically, echinochrome is closely related to vitamin K (which is 2-ethyl-3-phytyl-1:4-naphthoquinone), but it does not appear to possess any clotting activity.[429] The fact that echinoids have the best recorded clotting mechanism found in the phylum has already been referred to (Chapter 2), but a possible anti-haemorrhagic role for echinochrome does not appear to have been investigated. Some of the naphthoquinone pigment present in the amoebocytes exists in a colourless form, which upon oxidation turns red and eventually to a brownish-black,[370,522] suggesting that these substances could be intimately associated with the redox systems of the cells. It may be that by competing for the available oxygen, naphthoquinones could act as a controlling factor in the synthesis of another group of pigments which occur in echinoids and are also sensitive to redox potentials, namely the melanins. However, the interrelationships between these two pigments is still far from clear. The action spectrum for the spine-waving response in *Diadema antillarum* has a sharply defined peak at 455-460 mμ, and it is noteworthy that the spectral sensitivity of individual chromatophores involving pigment movement is also within this range, although the receptors involved in each case are different.[832,836] The λ_{max} of the acid form of the naphthoquinone pigment echinochrome A, which is widely distributed in this echinoid, is 463 mμ. Whilst it is not claimed that echinochrome A is involved in the photoreceptive reactions, the apparent coincidence of these figures is of interest. Recently, however, it has been shown that the absorption maxima of intracellular echinochrome is different from the extracted pigment, a value of 540 to 560 mμ being found.[531] Thus it seems unlikely that this pigment is involved in the primary photoreceptor mechanism. Much of the pigment in the nerves is in the amoebocytes, actual pigmented neurones being rare. Light is considered to have an inhibitory as well as an excitatory effect similar to that described for the siphons of *Spisula*.[395] In this preparation light of short wavelength had an inhibitory effect, possibly due to absorption by a carotenoid, and light of a longer wavelength had an excitatory effect, possibly due to absorption by a haem protein. Finally an algistatic function for these pigments has been suggested.[793b] Considerable progress has been made since the days of MacMunn, but so far as echinoderms are concerned, the crucial experiment has yet to be performed. As Fox said in 1953,[262] " . . . such is the disposition of echinochromes to which no definite physiological function seems yet to be assignable".

The final pigment of the echinoids to be considered is melanin. Just as in certain holothurians, the coelomic fluid of *D. antillarum* darkens upon exposure to air, passing through various intermediate colours to brown or black.[370,529] Microscopic examination reveals that the colour is due to the presence of spheroidal bodies within the amoebocytes constituting the clot, some of which may be in the process of disintegration. The resistance of the dark pigment to solvents, its bleaching by oxidizing agents and its reduction by ammoniacal silver nitrate indicate that, like the chromatophore pigment, it is melanin.

The darkening of the coagulum, a reaction which seems to have a pH optimum of between 6·5 and 7·0, is abolished by extremely low concentrations of cyanide (0·0003 M) and certain narcotics. This suggests an enzymically controlled oxidation. The coelomic fluid oxidizes several phenols including catechol, DOPA and tyrosine, but the use of isolated amoebocytes results in more rapid oxidation, suggesting that some inhibitory agent, which itself is found to be sensitive to methyl alcohol and may thus be a dehydrogenase, is present in the fluid component. Such was not found to be the case in those holothurians which were examined.[513,515,516] Finally, inhibition of the oxidation by hydrogen sulphide and by 0·01 M sodium azide enables the process of melanogenesis to be ascribed to an enzyme system with the properties of a phenolase rather than a cytochrome system. It seems certain too that the amoebocytes also contain the substrate for these reactions. A dual origin for the black pigments of echinoids has already been proposed (see above). Cytological work by these same authors has established that the melanic pigments are deposited in the test and epidermis by disintegrating amoebocytes. The exact region of deposition is narrowed by the introduction of iron saccharate into the coelomic cavity. Although an unnatural substance, it was phagocytosed by the amoebocytes and deposited in the superficial regions of the animal, especially in the actual chromatophores themselves. These facts, together with the observation that pigment charged epithelia can be eliminated in a manner suggestive of an excretory process,[519] render it unlikely that such pigment would ever be re-utilized by other amoebocytes.[154,155] More recent electron microscopical studies indicate that the granules come to lie in the intercellular spaces.[580] Continuous with these spaces are cavities scooped out of the cellular surface containing aggregations of pigment. A review of the pigmentary system in *Diadema* by Millott[524] indicates some interesting aspects regarding pigment deposition and turnover. Some pigments, it would seem, can be deposited at old sites to give a cast or chromatoglyph.

The melanins are undoubtedly protective in function, possibly screening a photolabile pigment.[521,775] Echinoids generally are negatively phototactic and their gross responses have been recorded on a number of occasions.[344,355,520,710,774,775] In tropical species where light intensities can reach high levels this phototaxis can take the form of diurnal migrations, sometimes into deeper waters or like *D. setosum* into crevices.[756] Positive phototaxis in response to artificial light has been recorded in *L. variegatus*, but even this animal becomes negatively phototactic in sunlight, executing its familiar covering reaction.[710] Similar positive phototaxis is also reported in *Psammechinus microtuberculatus*.[833] Obviously the physiological state of the animal will affect its response to illumination.

Physiological colour change in response to illumination has been recorded, for example, in *Arbacia lixula*[775] and *A. punctulata*,[408] the animals changing from black to brown in darkness. *Centrostephanus longispinus* changes from dark purple to a greyish colour in the dark and *D. antillarum* turns from an intense black to a greyish colour.[512,514] In this latter species, the diurnal colour change was maintained in constant darkness.[518] This capacity is lost in older animals due to the increasing deposition of melanin. Ambient light intensities also affect the rate of pigment accumulation (morphological colour change), due to the ability of the white spines to deposit melanin, but black spines are unable to resorb it.[423] Only a very small ray of light was required to cause pigment of a pale coloured animal (dark adapted) to disperse with the consequent darkening of the animal.[514] Ultra-violet light markedly increased the formation of melanin.[370] By using extremely small points of light only 3μ in diameter, Yoshida was able to prove conclusively that the chromatophores act as independent effectors and that their spectral sensitivity paralleled that of the whole animal.[831,836] The cellular mechanism governing the movement of the pigment granules is unknown.

Underlying the main layer of chromatophores is a discontinuous layer of colourless iridocytes, localized aggregations being known as iridophores. These are best seen in the dark adapted forms by reflected light when they appear as blue spots due to the scattering of light by their colloidal contents, which also consist of gelatinous plates. Aggregated into ridges and clumps mainly in the interambulacral region, they can take on the appearance of eyes and indeed were thought to be so by many earlier workers. Some have ascribed a light-emitting function to them, but most probably they function as a tapetum in increasing the sensitivity of the photoreceptive surface of the urchin, especially in fading light.[517,530] Additionally, electron microscopical studies of these structures indicate the presence of ribbon-shaped bodies composed of ellipsoidal structures.[389]

A reaction to light found in several shallow water echinoids is that termed the covering reaction. This appears to have been first described for the species *Toxopneustes* (=*Paracentrotus*) *lividus* as long ago as 1884. Accounts of this reaction have appeared since for a number of species including *C. longispinus* and *A. lixula*,[775] *Spherechinus granularis*,[777] *A. lixula*,[486] *Paracentrotus lividus*,[208] *L. variegatus*[521] and *Echinocyamus pusillus*.[576] This reaction assumes a similar pattern in each species. The tube feet and spines transfer material such as stones, shells, etc., from the substratum to the body of the animal, even floating debris being employed. In *L. variegatus* Millott has shown that a continuous light or changes in light intensity may initiate the response of the tube feet, which may at any time be overriden by a mechanical stimulus. The covering material is held in position by contraction of relays of tube feet without the participation of either the aboral or oral nerve rings. Diurnal covering occurs even in captivity, and activity is maximal about 3 hr after sunrise. The cause of this reaction has been variously attributed to changes in temperature, wave action, desiccation or the need for concealment. Originally it was suggested that it had a protective function for a photolabile pigment and recent work supports this contention. Although variations in colour within the population of *Lytechinus* did not result in a differential sensitivity in the urchins, covering once removed is replaced far more rapidly in the paler varieties. Injection of photosensitizing dyes such as Rose Bengal will cause urchins to cover in dim light, which suggests that the degree of protection afforded by the darker melanic pigments is insufficient in this situation.

Apart from the general negative phototaxis, the intensity of such responses as the covering reaction and physiological colour change increase with increasing illumination, as do movements of the tube feet, pedicellariae and spines. Thus when *A. lixula* is illuminated it will wave these appendages about, aimlessly at first and then they finally come to rest lowered on the illuminated side of the test and raised in the shaded region.[355,486] There can be no comparison of this reaction to the crossed reflexes of vertebrates, and going to other extreme Holmes compares the urchin to a *Volvox* colony.[355] von Uexkull[777] thought that this simple reflex did not involve the radial nerve, but Millott and Yoshida have contested this on very good evidence.[538] Nevertheless, it is difficult to reconcile this with Kinoshita's demonstration of chains of reflex relays within the integument of five species of echinoid.[404] It is to a sudden decrease in illumination that the best of these responses occurs, namely the shadow reflex. This reflex does involve the participation of the radial nerve cord and its sensitivity is about 1000 times greater than the response to an increase in illumination.

In a long series of papers this reaction has been intensively studied by Millott and his colleagues and as in their work on the chromatophores of *D. antillarum*, small spots of light were shone on to the surface of isolated pieces of test.[512,520,533,539,540,835,836] The amplitude, frequency and duration of the spinal response following extinction of the light was measured. The results indicate that all superficial regions and appendages of the body are sensitive to a decrease in the intensity of illumination with the exception of the spine shafts. Even the radial nerves, which are normally inside the test, are sensitive and where branches of these nerves emerge onto the surface along the ambulacral margins, the greatest degree of superficial sensitivity is to be found. The interambulacrum is the least sensitive, a gradation parallel to that for responses to increased illumination. In addition to the variation in equatorial sensitivity, there is some evidence of a polar one, although removal of the circumoral nerve ring did not affect the response. Removal of the radial nerve, however, abolishes it, but as mechanical stimulation of the spine is still effective it suggests that only the receptor pathway has been interrupted.

Readmission of light after a brief period of darkness was found to curtail or even abolish the expected spinal response. The degree of inhibition was found to be proportional to the intensity of the re-admitted light. By varying the intensity and area illuminated by the readmitted light, temporal and spatial summation over a wide area was found to occur. Millott's interpretation is that light *per se* has an inhibitory rather than an excitatory effect, which is therefore somewhat at variance with the more usual explanation. Removal of the inhibition, namely light, releases the characteristic spinal movements. A mathematical analysis demonstrated that for the closely related *D. setosum* the intensity I_1 and duration T_1 of the initial illumination was related to the intensity I_2 and duration T_2 of the readmitted light and to the interval of darkness t. The difference between I_1 and I_2 must always be a constant fraction of I_1 in order to effect a response and in this instance, therefore, the Weber–Fechner law may be said to hold, at least for moderate intensities. Keeping T_1 and T_2 constant it was found that the value for I_2 which would just inhibit a spinal response released by increasing periods of darkness for each of a series of different values of I_1 is, at first, constant. Eventually, however, a rapid rise in the intensity of the inhibiting stimulus is necessary. This is seen as the operation of at least two distinct processes during the period of darkness, and imbalance between peripheral excitation and inhibition at the level of the radial nerve cord, is thought to account for these results. Increasing temperature seems to result in increasing sensitivity to shading, at least at the higher values, but below about 12·5°C to 15°C this response breaks down. As mechanical stimuli still continue to be effective, such a breakdown most likely occurs in the receptor pathways and its exact nature is not known.

Analysis of the initial direction of the first movement of the primary spines when released from inhibition by shading showed that three directing influences were operative.[533] These were termed the surface factor, the ambulacral factor and the oral factor. The correspondence between the location of the stimulus relative to the spine and the number of initial movements in the direction of that stimulus was found to be high in the oral and adambulacral direction for the interambulacral spines. For ambulacral spines it was high in the oral and abambulacral directions. An anatomical basis for these results has been revealed by histological examination of the spinal innervation. Branches from the radial nerve cord, surfacing upon the test via the pore pairs, divide. One branch passes adambulacrally to supply the ambulacral spine and the other branch passes abambulacrally to supply the interambulacral spine. Thus on each side of a given ambulacral groove, the ambulacral and interambulacral spines are innervated in opposite directions. These nerves make contact with the nerve ring around the base of each spine and tend to enter it on the morphologically oral side. Thus the oral and ambulacral factors are to be seen as representing differing aspects of the same thing. The sphere of influence of each radial nerve has been shown to extend beyond the confines of its own ambulacrum in *Anthocidaris crassispina.*[404] Further evidence to support this has been provided by experiments involving the illumination of a denervated ambulacrum. This results in the interambulacral spines pointing to the nearest intact ambulacrum, and indeed it has been established that the field of influence of an intact radial nerve may extend halfway around the test. When more localized responses are involved, the surface factor, which of course represents conduction through the superficial nervous system, may have an overriding effect, but the more widespread responses involve the nerve cords.

Proprioceptive spinal relays of the kind demonstrated by Kinoshita in *A. crassispina* as a result of chemical stimulation do not appear to occur in *D. antillarum* as a result of photic stimulation;[404] neither does the "freezing" of the spines so widely reported for other forms of stimulus. It is evident that photic stimulation, involving as it does the radial nerves, elicits a more complex train of events than do other stimuli and the observed "on" and "off" reactions are reminiscent of the responses obtained from sophisticated photoreceptors and nervous systems. This comparison is further strengthened by detection of "on" and "off" responses electrophysiologically from the radial nerve cord of *D. setosum* when stimulated by a light spot.[748] In no sense, therefore, should this highly complex dermal light sense be regarded as primitive.[526]

When stimulated, ambulacral spines oscillate at a higher frequency than interambulacral ones, even if they happen to be at a greater distance from the point of stimulus. Information from the effects of temperature upon the various phases of these reactions has led to the suggestion that the radial nerves act as pacemakers as well as initiators in this spinal waving. The differing speeds of conduction in the superficial nervous system, in the efferent pathways about 60 mm/sec and in the afferent pathways about 22 mm/sec, has led Millott to suppose that the nervous system is at least in part organized into tracts. This suggestion, based upon mainly physiological evidence, is in accord with that demonstrated by Kinoshita[404] and by Smith[722] from his studies of asteroids. Further reviews of this aspect have been published[99,225,725] and further consideration is given to the nervous system in Chapters 8 and 11. The apparently simple behaviour of the echinoderm is deceptive and in much recent work the "central nervous" role of the radial nerve cords is being emphasized, rather than the more classic comparison to the more diffuse nerve net of the coeloenterates.

Asteroidea

Whereas in echinoids quite a diversity of pigment types is to be found, the asteroids have a more restricted range. Carotenoids, together with their conjugation products with proteins, seem to be largely relied upon to supply the colour variations within the class; nevertheless, these can be quite striking. In many respects, they exhibit some similarities with the crustaceans, a point often noted but obviously of no phyletic significance.[1] The earlier workers in this field reported the presence of a number of lipochromes with one or two absorption bands.[425,569] Additionally there were those soluble pigments which changed colour upon heating or treatment with alcohol. These latter were undoubtedly carotenoproteins and are usually red or purple in colour. Since those times numerous workers have directed their attention to qualitative studies of echinoderm and particularly asteroid carotenoids.[1,262,263,473–476,585,589,770,791,795] Briefly, the main outcome of this work is that a wide range of carotenoids are common in the Asteroidea including β-carotene, xanthophylls, both free and esterified, of which the most frequently occurring include astaxanthin, lutein, and certain keto-carotenoids. The form of the pigments may change within a given species with age;[585] in the young *Echinaster sepositus*, for example, it was mostly free astaxanthin; in the older specimens it is esterified.[585] These pigments are most abundant in the integument, gonads and hæpatic caeca, but do not appear to be localized in chromatophores. The first to recognize the nature of the carotenoproteins appears to have been Verne (1921, quoted Abeloos[1]) who termed them carotenalbumines, although their occurrence had been noted earlier. The various purple and violet colours of several asteroids were ascribed to these complexes but they were not characterized further. Similarly a blue pigment was isolated from the integument of *Asterias rubens* and a violet/red one from *Porania pulvillus*.[223] The red carotenoid component of the chromoprotein from *Asterias* was named asteric acid, but was almost certainly astaxanthin.[386] This substance, also conjugated with a protein, was also isolated by Vevers from the same animal and he showed that it was responsible for the brownish tints that this animal exhibits.[791] Astaxanthin is readily oxidized to astacene *in vitro* and this reaction may account for the supposed existence of astacene *in vivo*. The blue shafts of the spines of *Astrometis sertulifera* are easily reddened by reagents denaturing proteins, as also is the integument of the blue and purple starfish *Pisaster giganteus*. A carotenoprotein is therefore most likely responsible for these colours.

Whereas morphological change, particularly in relation to advancing age, has been recorded from several echinoids, no similar information is available for asteroids. The report that physiological colour changes occur in *Marthasterias glacialis* in response to injection of nerve extracts is extremely interesting.[780] Two components were identified, one causing blanching of the integument and the other a darkening. These were not the only effects of this extract, and it will be considered again in Chapter 10. The origins of many of these carotenoids is not clear, but some at least have a dietary source. In the herbivorous echinoids and holothurians, a concentration of about 0·5 mg/100 g fresh weight of carotenoid has been recorded, of which some 54% is due to xanthophylls.[263] In the carnivorous asteroids and ophiuroids a total body concentration in excess of 1·6 mg/100 g fresh weight is recorded, of which more than 90% is due to xanthophylls. Animal food is richer in xanthophylls and the carnivores seem to be more able to store carotenoids, although these differences are more probably metabolic than dietary. Antheroxanthin, metridioxanthin, zeaxanthin and diatoxanthin in four species of asteroid as well as several unnamed compounds are also recorded. A free carotenoid acid with absorption peaks similar to those for mytiloxanthin had been found in the starfish *P. ochraceous*.[262] Mytiloxanthin was first isolated from the mussel *Mytilus californianus*, upon which this starfish preys.[699] In the mussel there was evidence of a metabolic link with zeaxanthin. There were numerous carotenoids from the Mediterranean asteroids *Asterina panceri*, *Ophidiaster ophidianus*, *Astropecten aurantiacus* and *Marthasterias glacialis*.[581–584] These included neo-β-carotene, γ-carotene, cryptoxanthin, zeaxanthin, pentaxanthin, lutein and a whole range of hydroxy- and keto-derivatives including 4-keto-β-carotene, which is also known as echinenone. This substance which occurs most commonly in the echinoids, from which it was first isolated by Lederer,[455] has in addition to the species mentioned above, been recorded from *E. sepositus*.[585] It is the only animal carotenoid so far recorded as possessing pro-vitamin A activity. The biosynthetic pathways for the conversion of dietary β-carotene into astaxanthin, involves several of the carotenoids as intermediates, including, in the case of *Asterina panceri*, echinenone. Thus a possible function has been found for this rather enigmatic substance, for, although echinenone has been shown to have pro-vitamin A activity, in other animals[460] no vitamin A has yet been detected with any certainty in asteroids.[669,795]

In contrast to the large amount of work carried out upon the photosensitivity of echinoids, little has been attempted on asteroids, other than fairly simple experiments designed to investigate general phototactic behaviour. On this basis, asteroids can be roughly divided into three groups, although it should be noted that such behaviour can be modified by other factors such as hunger or age. The first category refers to those species which are negatively phototactic, such as *Astro-*

metis sertulifera. This species is aroused by light, settling down once again when shade was encountered[371] and strictly such behaviour should be called orthokinesis. Permanent illumination results in the animal coming to rest with its oral structures protected. Some other species, including *Asterias forbesi* and *Henricia* (= *Echinaster*, = *Cribrella*) *sanguinolenta*, would withdraw from bright or sudden illumination,[562,563] or in the case of *Astropecten aurantiacus* the tube feet retract and the ambulacral grooves close.[344] Similar results were obtained with *Asterias forbesi*[547] and with *Oreaster reticulatus*,[163,164] a sudden shadow being effective in eliciting these reactions. In view of Millott's findings with echinoids, this latter example could well bear reinvestigation. Secondly comes those asteroids which are positively phototactic, and this seems to include the greater proportion of species. As early as 1881, Romanes and Ewart found *A. rubens*, *Solaster* (= *Crossaster*) *papposus* and *Astropecten aurantiacus* to be strongly photopositive.[676] Three further species included *Asterina gibbosa*[640] and *E. crassispina*[161–165] which would even select the lightest of three backgrounds.[640] Finally *A. gibbosa* and *A. panceri* fall into a third category, avoiding extremes of light and shade.[486] Only young specimens of *A. gibbosa* were photopositive to light in a directional fashion as well as to the background.[378]

Asterias rubens can fall into any one of these categories, as most probably can a number of species depending upon other intrinsic factors. It may be generally negative to light,[66] positive to weak light[804] or positive to light of all intensities.[630] Just[377] described an interesting phototactic response which led the animal to move along a pathway which was the resultant of two incident light beams. This experiment has been criticized upon seemingly reasonable grounds,[199] but the significant observation was made that the isolated arm of a normally photopositive animal will travel with the base of the arm towards the light source with the tip of the arm which bears the eyespot, recurved over the aboral surface so as to face the incident light beam. An interesting response to light is also shown by *Astropecten polyacanthus* which emerges morning and evening when the light intensity reaches a certain value, although this value was unfortunately not determined.[553] This rhythm was maintained for two to three days in absolute darkness and its vigour was accentuated by hunger. A diurnal response to light was also found in the young individuals of *Asterias forbesi* which were positively phototactic during normal daylight hours and negatively phototactic during the hours of darkness.[672] The most effective portion of the spectrum for eliciting this response was around 450 mμ. Green light was found to attract the cushion star *Pteraster tesselatus*, but blue and yellow light repelled it.[674] Individuals of *Asterias* sp. collected at the end of an aquarium illuminated with a red darkroom lamp.[624] The absence of any intensity measurements prevent any conclusions from being drawn from these observations.

Whilst most authors agree that the general body surface of asteroids is sensitive to light, van Weel going as far as to propose that the skin could mediate a shadow reaction,[804] they are by no means in accord over the role of the terminal optic cushions. Some authors claim that a directed response is abolished by extirpation of the optic cushion,[199,377,378,630,640,675] whilst others have found that the removal of the optic cushion did not affect the animal's responses.[165,486,563,667] However, it should be pointed out that experiments where a choice of illuminated or shaded regions of an aquarium tank is offered would be just as successful in animals lacking a directional photoreceptor provided that the general body surface were sensitive to light. Recently it has been observed that the tips of isolated arms of *A. amurensis* move downwards when illuminated and upwards when the light is shut off,[837] the reaction time being between 4 and 7 sec. Removal of the ocellus does not abolish this response, but it raises the threshold by a factor of about 10. There is, however, a shift in spectral sensitivity upon extirpation of the eyespots from 485 mμ to 504 mμ. A considerable amount of a soluble red pigment with an absorption maximum of 480 mμ has been extracted, yielding both epiphasic and hypophasic carotenoids upon denaturation.

Being the more obvious structures, morphologically speaking, associated with vision, the pigments extracted from the optic cushions of *M. glacialis* have been examined.[534] Not surprisingly these consisted of two pigments already known to occur commonly in the integument, namely, β-carotene and esterified astaxanthin, this latter pigment being restricted to the individual optic cups giving them a deep orange colour. The pigment cells of the optic cups are slender and prolonged at one end with the pigment distributed more or less continuously throughout the cytoplasm. However, recent work on the fine structure of the optic cushion has clarified the relationships of the various cell types to be found there. A layer of cells 1 μ thick inside the optic cup with inwardly

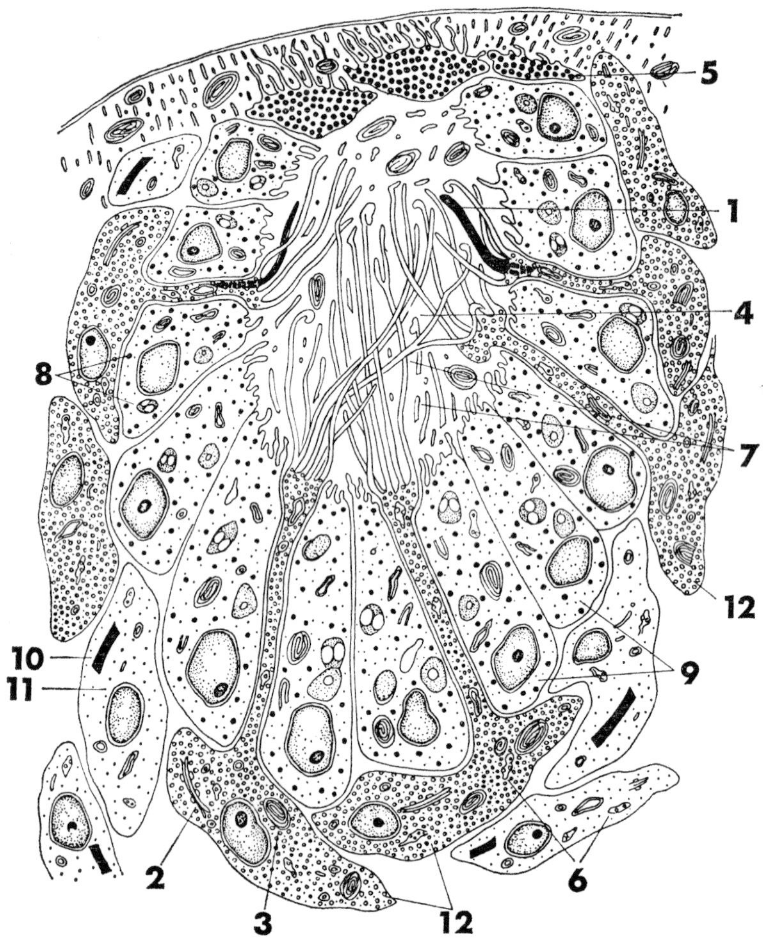

FIG. 14. Diagram of the eyespot of *Asterias rubens L.* 1, cilium; 2, Golgi apparatus; 3, numerous lamellate bodies; 4, lumen; 5, "lens"; 6, mitochondria; 7, microvilli; 8, pigment granules; 9, pigment cells; 10, supporting fibres; 11, supporting cells; 12, sensory cells.

directed microvilli is reported,[625] which may be a lens-like invagination. Separate pigment cells and photoreceptors are distinguishable in *A. rubens* eye cups.[786] The photoreceptors underlie the pigment cells and send processes in between the latter which then divide up in the lumen of the optic cup to give rise to the microvilli. Each process also bears a cilium with typical rooting structure. The cytoplasm of the photoreceptors contains Golgi apparatus, tubular mitochondria, and many pale regions which are continued into the processes between the pigment cells but not into the microvilli. The pigment cells are also furnished with Golgi apparatus and mitochondria. The pigment is overlain with an assortment of granules and globules. The eye cups are cut off from the cuticle by special cells suggestive of a lens. Comparisons have been made between the optic cup of *Asterias* and the ocellus of the hydromedusan *Polyorchis* except that in this coelenterate the microvilli arise from the cytoplasmic covering of the cilium instead of cellular processes. The ciliary nature of the photoreceptor apparatus has been confirmed, but the surface membrane of the cornea had long microvilli projecting into a layer of jelly in several species of asteroid which were latterly investigated.[214] The evidence for these pigments acting in a photochemical capacity is slender and their function may be solely that of a screen, for astaxanthin and other carotenoids carry out this function for the retinal cells of Crustacea.[800] Some bleaching in methanol solution has been observed, but this was too slow to be of any photochemical importance.[534] A pigment from the dark adapted eye of *A. rubens*, which had an absorption maximum at 480 mμ, also exhibited bleaching. The optical density decreased markedly between 500 and 600 mμ, but here again such bleaching was too slow to be of value in a visual process, unless such an energy transfer did not involve extensive photodecomposition.[624] An unidentified pigment extracted from both the skin and eyespots of *A. forbesi* has a sharp absorption peak at 320 mμ and another less well-defined one at 460 mμ to 500 mμ.[667] Bleaching occurred but slowly in both these extracts. An unusual pigment with an absorption maximum at 495 mμ, similar to that for astaxanthin, has also been extracted from the same species.[667,671] This new pigment exhibited rather anomalous behaviour upon bleaching. When light of wavelength below 450 mμ was used, there was a maximal drop in the optical density at 585 mμ, bleaching at 485 mμ being minimal. Using light of a longer wavelength, between 600 and 700 mμ, there was maximal bleaching at 485 mμ and minimal bleaching at 580 mμ. There is some suggestion that a regenerative process might be at work and at the longer wavelengths its absorption and bleaching characteristics were similar to those of the pigment from *A. rubens*.[624] The difficulty of separating off the protein moiety of this pigment suggested that the photosensitive reaction did not involve a "retinene-like" splitting, although a Carr–Price test for vitamin A on material extracted from dark adapted eyes gave a positive result. If substantiated, this is the first report of the occurrence of vitamin A in the echinoderms. More recent work, using a modified extraction procedure, has yielded a violet photosensitive pigment with an absorption maximum at 523 mμ which is rapidly converted to a peach coloured pigment with an absorption maximum at 490 mμ. This pigment has been named stellarin.[668,670]

The exact mechanism linking the light stimulus to the generation of a nerve impulse is far from being understood, but almost certainly a photochemical reaction involving some pigment is indicated. Whilst not of the vertebrate rhodopsin pattern, some carotenoid may well be implicated. The situation is made still further interesting by the development of a potential difference across the optic cushion of *Asterias* sp. when this is illuminated, the corneal surface becoming negative.[325] Unlike most other retinograms and the electrical activity recorded from the radial nerve of *D. setosum* in response to illumination, there was no "off" response. However, a slow fall in electronegativity during the period of illumination suggested some form of adaptation. The existence in the skin of similar carotenoids to those found in the optic cushion takes on added significance in the elucidation of the photosensory mechanism of this group of animals.

Finally the pigments possessing a cyclic tetrapyrrole structure, the porphyrins. Although MacMunn[568] claimed the presence of haematoporphyrin in the integument of *A. rubens*, it is now known that the molecule lacked the iron atom and was in fact protoporphyrin.[397] In a more comprehensive survey this same substance was also found to occur in *Luidia ciliaris* and *Astropecten irregularis*[398] together with chlorocruoroporphyrin. It was the first time that the free substance had been found. Usually it is combined with iron and a protein to form chlorocruorin, the familiar green respiratory pigment of sabellid blood. Porphyrins have been shown to occur frequently in association with calcium in many ossified skeletal structures such as mollusc shells, teeth and bones, especially in certain pathological conditions. Its occurrence in the calcified integument of these animals is therefore interesting, although it is difficult to determine whether it is actually associated with the spicules. The distribution of protoporphyrin is erratic, however, for it does occur in the very closely related *M. glacialis* which has an almost identical food to *Asterias* and the larvae of the two species are virtually indistinguishable. The absence of protoporphyrin and its precursors from the prey of the starfish indicates that its presence is the result of some metabolic process and its formation from the "enterochlorophylls" or phaeophorbides, which are plentiful in the gut caeca, seems a likely one.[398] The concentration of protoporphyrin in the integument of *A. rubens* is low in the pale specimens and increases in the darker brown and violet ones, so that they can be arranged in a sequence. Protoporphyrin does not contribute to the coloration of these animals and its function is unknown.

Ophiuroidea

In the ophiuroids the red and yellow colours are also due to carotenoids, the presence of lutein in four species of ophiuroid from the Swedish coast being recorded.[473,476] Three species from California were examined with basically similar results, namely, that the oxygen-containing pigments were predominant.[262,263] These included dinoxanthin (= taraxanthin), pectenoxanthin, together with various esters and several unnamed xanthophylls which seem to be peculiar to the ophiuroids. Indeed each class of echinoderms appears to possess certain carotenoids unique to itself.[262] Unlike the asteroids, however, these carotenoids occur in specialized chromatophores called lipocytes in *Ophiocomina nigra*.[257] These dendritic cells lie beneath the melanocytes and are not syncytial. The pigment within these cells may be dispersed or granular and is probably combined with protein, further, their chemical nature has been confirmed adding β-carotene and an acidic carotenoid similar to mytiloxanthin to the list, but lutein was not recorded. There is no evidence of any pigment migration under varying conditions of light. There would seem to be a geographical factor in the occurrence of the various colour variants together with a qualitative difference in the proportions of the pigments.[255,257] Carotenoid-containing cells in localized patches also occur on each vertebral ossicle. These cells are unlike the lipocytes already mentioned, being drawn out into tail-like processes at one end, somewhat after the fashion of the cells of the optic cups of *M. glacialis*.[534] Chromatographic analysis showed them to contain similar pigments, β-carotene and an ester of a mytiloxanthin-like carotenoid instead of the more usual esterified astaxanthin. The pigment in these spots fades rapidly when exposed to bright light which may be the result of either bleaching or oxidation or both. The function of these vertebral pigment spots is unknown, but being situated on the aboral side of the vertebral ossicle and being covered by the upper arm plate only, could suggest a photoreceptive role, but definite evidence is lacking.

Also present in the integument of *O. nigra* are pigment-containing cells whose dendritic processes appear to fuse to give rise to a syncytial layer which would seem to be unique in the echinoderms.[256] The pigment they contain has been shown to be a true melanin and they have been designated melanocytes, and contrary to the echinoids the pigment does not undergo any form of migration in response to illumination. Also associated with these melanocytes are three fluorescent pigments, one of which is riboflavin, the other two being possibly pteridines. Non-melanic ophiuroids contain only a trace concentration of these fluorescent substances, although the enzyme system of the phenolase type capable of the production of melanin is still present and the *in vitro* formation of the pigment from these forms has been achieved. The regenerating tips of the arms of non-melanic brittle stars are much darker then the remainder of the body. Obviously some inhibitor of melanogenesis must be present in non-melanic animals which is detroyed both by the process of wound healing and by the conditions of the *in vitro* experiment. An inhibitor of melanogenesis with some properties of a dehydrogenase was present in the coelomic fluid of the echinoid *D. antillarum*.[529] Reduced melanin is fawn in colour and fawn areas of otherwise dark animals have larger amounts of pteridines. There is evidence to show that in other phyla pteridines suppress melanogenesis and that riboflavin accelerates it. Thus the balance in concentration of these two substances, each of which may be closely associated with the maintenance of the correct redox potential, could be invoked as a controlling mechanism of the process of melanogenesis rendering it possible for any shade from brown or fawn to black to occur. Such a situation closely parallels the condition in the echinoids, except that the redox balance there may be controlled by quinones. Melanin acts as an extremely good protective screen and there is some suggestion of a selective advantage in the possession of this pigment, for the lighter coloured animals are more frequent in deeper waters.[255] The colour of the substratum may also not be without effect upon their coloration.

Many workers have commented upon the general negative phototaxis exhibited by ophiuroids.[163,486,502] Like many asteroids they will seek out the protection of stones, crevices and even the darkened walls of an aquarium. The arms, especially the aboral surface, appear to be the most sensitive, possibly for those reasons mentioned above. Removal of any protective covering will result in the animal moving off in the direction in which it was removed, often without any reference to the direction of the incident light, suggesting that it retained some "memory of the past stimulation."[163]

LUMINESCENCE

Luminescence, so often associated with deep sea life, is only to be found in this last group of echinoderms to be considered, namely the ophiuroids. The source of this luminescence appears to be the unicellular glands found in the spines and on certain parts of the body, namely, the lateral shields and possibly the tube feet, although this latter instance has not been confirmed.[183,329] Many non-luminescent ophiuroids have similar glands within their spines and both types have acid mucopolysaccharide inclusions.[94] Upon either mechanical or chemical stimulation these species emit a greenish-yellow luminescence which appears to be under nervous control. A wave of luminescence travelling from the point of stimulation will not pass across a transected nerve. No luminescent secretion has, however, been obtained from either extracts of glands or whole animals and there is as yet no indication of the presence of a luciferin/luciferase system.

It is almost impossible to summarize the contents and notions presented in a chapter of this length and diversity, but perhaps a few general observations would not be out of place. One of the most significant facts in connection with the effects of mechanical stimuli is the superficially simple behaviour of the seemingly independent effector organs such as the tube feet. Their overriding positive stereotropism is largely the basis for the initiation of the righting reaction, so common in this phylum. But their obvious co-ordination during this reaction highlights the more central control either originating in or mediated by the radial nerve cords. A similar viewpoint is taken by Millot who considered quite different modes of stimulation. The holothurians exhibit a quite definite bilateral symmetry, and in some of their activities other groups too show some tendencies in this direction. One does not have to look far in the animal kingdom to see what advances are associated with such a development.

The graded response, familiar in coelenterates, is seen here not only in the behaviour of the surface appendages but is reflected also in the behaviour of the whole animal. Raising the stimulus intensity above a certain threshold value will often result in the reversal of the response. The exact nature of the stimulus may be relatively unimportant, and as a consequence earlier workers subscribed to the existence of a "universal sense organ", especially in the absence of any great range of clearly differentiated sense organs. The pigmentation of echinoderms is extremely variable and each class seems to have its own characteristic complement. True physiological colour change is rare and has been recorded from a few echinoids only. Possibly the simplest array of pigments is to be found in the asteroids which utilize the carotenoids almost exclusively, and their conjugation products with proteins, thus showing parallels with the Crustacea, especially in regard to visual processes. Holothurians and ophiuroids possess carotenoids and melanin and in the stellaroid groups at least there is some evidence of a photoreceptive role for them. Additionally the crinoids possess traces of the aromatic naphthoquinone and anthraquinone pigments which are of an especial palaeontological interest. Although the echinoderms have arrived at the organ level of organization, in many cases function is only physiologically differentiated within the cell which lacks to any great extent any morphological differentiation. In some ways such a condition is reminiscent of protozoan organization, but with added complexity. The reception of light by echinoids, for instance, where even the nerves are light-sensitive, serves to demonstrate the complex interplay of excitation and inhibition, of peripheral and central control, which is hidden by a superficial simplicity and has led in this instance to a revaluation of the part played by the nervous system. Such a concept is far removed from von Uexkull's "Reflexrepublik" and reflects to some extent a "hierarchical" approach. In this light much of the early experimental work upon the behaviour of echinoderms should be viewed and possibly re-examined.

CHAPTER 8

PHYSIOLOGY OF THE WATER VASCULAR SYSTEM AND THE NEURAL CONTROL OF LOCOMOTION

Morphology

Probably the most characteristic feature of echinoderms is the presence of a water vascular system. Although variations occur from class to class, the basic organization is similar throughout and the reader is referred to almost any of the standard zoological texts for the anatomical details and their modifications. Briefly, however, the system consists of a water vessel passing along each radius which may of course become subdivided several times in those animals with numerous arms. All the vessels are united at their proximal ends by a circumoral water vessel from which may arise one or more stone canals. This canal, which may or may not be calcified, passes aborally in an interradial position and opens either on to the surface of the animal by way of a calcerous plate called the madreporite or alternatively the madreporite may be internal as it is in the vast majority of adult holothurians. From the radial vessels numerous lateral canals pass to the tube foot ampulla systems, in most animals the entry into them being guarded by a flap-like valve, the function of this valve being to prevent the backflow of fluid into the radial vessels during the protraction phase of the tube feet. In some groups, such as crinoids and ophiuroids, specialized ampullae are absent and muscular dilations of the radial vessel or of a special "head" region of the tube foot serve a similar function.[95,577,813] In many holothurians tube feet have become scattered over the surface of the body, with the consequent loss of the more obvious signs of pentaradiate symmetry. These podia are supplied by elongated lateral canals called podial canals which may be branched and follow an erratic course. The extreme ends of the radial vessel terminate upon the surface of the body in a small tentacle.

Various other structures are more or less intimately associated with the water vascular system, including the axial sinus. In various species of echinoid, asteroid and ophiuroid this has either an opening into the top of the stone canal or madreporic ampulla or is confluent with the stone canal at some point along its length. The occurrence of this junction has been denied by some authors,[108] but is confirmed by many others.[79,109,179,290,335] When present, the Polian vesicles open off the circumoral water ring and can vary in number as do the Tiedemann's bodies, which are small lymphoid structures seemingly restricted to asteroids. Either small protrusions or spongy bodies, which are thought to be homologous with Tiedmann's bodies, are associated with the

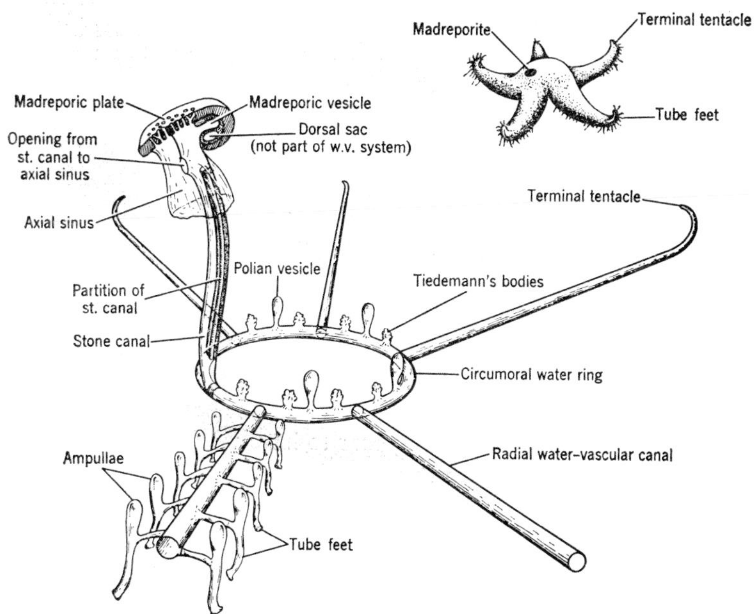

FIG. 15. Diagrammtic representation of the asteroid water vascular system, drawn as though the walls of the vessels are transparent. Tube feet are drawn on one arm only. Arrangement of Polian vesicles and Tiedemann's bodies generalized. Inset: those parts of the system visible on the exterior, drawn in the same orientation.

circumoral water ring in echinoids and ophiuroids. They are thought to play some part in the production of the amoebocytes so common in the water vascular fluid (see Chapter 2).

In the physiological discussion which follows, the main description of the function of the various structures of the water vascular system is based largely upon the asteroids, for in this group the system has received much attention. However, there is no reason for believing that it differs greatly from that of the remaining groups. Traditionally it has always been supposed that the water vascular system was filled with sea water and further, that any losses occurring from the functioning of the tube foot ampulla systems were made good by entry of water through the madreporite. Such a function for the madreporite was current at the time of Hartog[328] or even earlier and was without any real experimental evidence other than that of ciliation. It has been reiterated ever since in numerous zoological texts,[448] although lately some doubt has been cast upon the idea, again without any experimental evidence.[362] It would thus seem logical to start with a consideration of the madreporite.

Water Currents

The asteroid madreporite has an irregular furrow upon its surface which is lined with flagellated columnar cells. These beat in a variety of directions, being centrifugal in *Solaster* and *Porania*[291] and *Asterias forbesi*[97] but more tangential in *A. rubens*.[290] The main purpose of these currents would seem to be to keep the madreporite clear of obstruction. The pore canals which communicate between the bottom of this furrow and the top of the stone canal appear to have weak flagella beating in an outwards direction. With the exception of a few elasipods, all adult holothurians have an intracoelomic madreporic body which is also perforated and may even possess a vestigial furrow. Numerous stone canals may also be present, as there are in crinoids, but this group lacks a madreporite. Many species of ophiuriod too have multiple stone canals and madreporites, but these open onto the body surface in the oral position.

The direction of water currents within the stone canal itself has been debated for some time and it is not impossible that the direction of beat can on occasions reverse, although direct observations of such behaviour are lacking. Hartog[328] thought that there was an outward current of fluid through the madreporite and that this system represented a nephridial relic. Other workers have thought that such outward currents were artifacts and that they were the result of the activity of the superficial cilia and that the fluid movement was in fact inwards.[174,175,478] Such observations were extended to include certain species of holothurian and echinoid,[561] and this author also mentioned the flap valves guarding the entrance to the tube foot ampulla systems which were first described by Lange[446] and around which now centres much of our ideas as to the way in which the tube feet function. Other evidence by a variety of workers demonstrates pretty conclusively that the main direction of ciliary movement along the stone canal was in an oral or inwards direction.[37,97,290,367,646a] In most instances, however, excised preparations were employed and it may well be that under these circumstances the apparent rate of movement of particles was greater than in intact animals, where the fluid in the water vascular system will be under some pressure. The stone canal often has a scroll-like ridge running along its length and the suggestion has been made by Bamber that the ciliation in this region is in the opposite direction, supposedly imparting some sort of fluid circulation.[37] Confirmatory evidence is, however, wanting, but she came nearer than her contemporaries did in elucidating the function of this structure in echinoids. She proposed that the action of the madreporite was to maintain a balance between the internal and external pressures rather than to maintain a hydrostatic pressure within the water vascular system itself.

The Hydraulic Nature of Tube Foot Operation

The oral end of the stone canal opens into the circumoral water vessel from which arises the numerous appendages already mentioned, and in certain cases, notably the ophiuroids, the buccal tube feet which consequently lack ampullae. The radial water vessels also take origin from this ring and these pass along the radii eventually to supply the tube foot ampulla systems. In most groups of echinoderms the interior surface of the entire water vascular system is ciliated or more strictly flagellated throughout and thus the movement of the fluid is facilitated. The flap valve at the entrance to the tube foot ampulla systems is lacking in crinoids. The interdependence of these systems during the protraction and retraction phases of the stepping cycle has been recognized for some time, but the exact nature of the mechanism involved has been less certain.[484,610,676] The observation by Smith in 1946 of the constancy of podial diameter during this cycle, resulting from the presence of connective tissue rings arranged at right angles to the major axis of the podium and situated between the epithelial and endothelial layers, demonstrated the anatomical basis crucial to any hydraulic theory of operation.[720] The contraction of the ampullary muscles with the consequent flow of fluid into the podium leads to its passive extension, expansion being impossible. Retraction of the podium results from contraction of the longitudinally arranged musculature of the podium, the fluid passing back into the ampulla whose muscles are relaxed. Under these conditions it is supposed that the hydrostatic pressure within the system would keep the flap valve shut and prevent any fluid transfer to the radial water vessel. Hydraulically, therefore, the tube foot ampulla system is a closed one. A certain amount of musculature is, however, associated with the valve and it seems possible that it could on occasions be held open with the consequent deflation of the system. Mangold noted in 1908 that rarely in isolated preparations from *Luidia ciliaris*, both tube foot and ampulla contracted simultaneously.[484] The ophiuroids are generally considered to have more limited powers of protraction. An ampulla as such is absent, but a specialized region of the tube foot serves a similar function. Accessory vesicles or portions of the water vascular canal which can be separated by serially arranged sphincter muscles increase the volume of the reservoir. Several different modes of operation are possible for this system.[813] Firstly, fluid can pass in and out of the radial canal with movements of the tube feet in a manner analogous to that in asteroids. Secondly two methods of isometric extension can be performed with the flap valve shut at all times. This is due to the reciprocal activitities of the bulb and stem segments of the tube foot. The podium is equipped with two adjacent but separate layers of spirally wound connective tissue fibres, one a left-handed and the other a right-handed helix. In the Crinoidea, however, not even the valves are present and successive portions of the radial water vessel can be isolated to act as reservoirs whose dimensions can be increased by contraction of intrinsic muscle strands.[577]

Water Loss and Tube Foot Function

It has always been assumed that as a result of the operation of the tube feet, water was lost from the system by what was usually ascribed to diffusion.[448] In view of the part played by hydrostatic pressure such loss, if indeed it does occur, should be more accurately described as ultrafiltration. Until recently no evidence was available to support this contention and it is still unknown whether ions as well as water are lost during the stepping cycle of the tube feet. After the stepping cycle there follows an inactive phase which may be more or less prolonged. Hence not all the tube feet possessed by an animal are actively engaged in supporting it at any one time. In fact *A. rubens* has approximately ten times the number of tube feet that it requires for actual support.[399] It follows that the load a tube foot can support is directly related to the internal hydrostatic pressure, for there is no other skeletal support. The value for this load for an average podium of *A. rubens* was determined by Kerkut[399] using an apparatus resembling a torsion balance, and was about 40 mg. Hydrostatic pressure measurements within the water vascular system have been virtually restricted to the larger tentacular ducts of holothurians. Pressures of between 13 and 25 cm water have been recorded in *Caudina chilensis*[829] and similar values from *Thyone briareus*.[642] Bearing in mind the very small volume of the tube foot ampulla systems, it is desirable that in order to record maximal pressures occurring, the recording instrument should be as isometric as possible in its function. Using a spoon gauge,[627] pressures within the tube feet of *A. rubens* were found to range from 20 to 40 cm water during the contraction of the ampulla.[62] By making measurements of the internal diameter of the podia, it is possible to calculate that an average tube foot from this animal with an internal pressure of 30 cm water will generate a thrust of some 42 mg. This figure closely corresponds with Kerkut's experimentally determined value, and one which suggests that further support from other elements is unlikely and unnecessary. More recent determinations of the podial hydrostatic pressure in echiniods indicates a slightly lower figure of around 20 cm water.[233]

Permeability

The walls of the tube feet have been shown to be very permeable to water, oxygen and ions (see Chapters 2 and 3). It is therefore to be expected that under the influence of a measurable hydrostatic pressure, water and possibly ions too will be lost to the external medium. In view of their varying size during the protraction and retraction phases of the stepping cycle, the varying numbers in operation at any one time and the variation in the hydrostatic pressure during the operation of such cycles, a direct measure of the water loss from the water vascular system would be extremely difficult. Measurement of the water loss due to ultrafiltration from single isolated podia have, however, been made.[62] By making assumptions in regard to the variable parameters outlined above, an approximate value for the permeability of the tube foot membrane can be made. It is given below, together with two examples taken from Prosser and Brown:[642]

Necturus kidney tubule: $15 \times 10^{-9} cm^3/cm^2/sec/cm\ H_2O$
A. rubens tube foot: $25{\cdot}5 \times 10^{-9}$
Frog mesenteric capillary: 560×10^{-9}

A driving force of around 30 cm water is of a similar order of magnitude to that available in the mammalian kidney for filtration. The quantity of fluid passing across the glomerular membrane is, however, some twenty times greater than that passing across a similar area of tube foot wall. This is not surprising, for the glomerular membrane is specially constructed so as to permit a high filtration rate. From Kerkut's data, the number of tube feet in the fully protracted state (i.e. supporting the weight of the animal) is known and an estimate of the total surface area involved can thus be made. For a 50 g starfish a figure of 17·6 cm^2 has been arrived at and substitution in the above formula shows that such an animal should be losing water from the water vascular system at the rate of some 0·5 ml/hr. Thus the classical idea that the water vascular system loses water as a result of the operation of the tube feet has been substantiated. Such a quantity of fluid should therefore pass in through the madreporite, but whether the equally classic idea that this water loss is made good by such an inward movement must now be examined.

Water Replenishment

Smith[720] has pointed out that the volume of the water vessels is only some 1 to 2% of the total volume of the water vascular system. Thus whilst they are entirely inadequate to act as reservoirs for the operation of the tube feet, they could well act as the route for conducting sea water to those systems to replace that lost by ultrafiltration. As already described, there is strong evidence of an orally directed current in the stone canal. There is, however, very little evidence to show that liquid does pass into the stone canal from the outside medium,[290] and none whatsoever to suggest that any fluid from the stone canal actually finds its way into the radial water vessel and tube foot ampulla systems. In fact evidence will be presented here to suggest that the topping up of these systems is by an entirely different method. Many workers have tried keeping echinoderms in water in which carmine or other finely divided material such as indian ink have been suspended. The superficial ciliation in these cases is usually sufficient to prevent such particles from alighting upon the surface of the madreporite. Even after removal of the surface layers of the madreporite no penetration of the stone canal has been observed. Colouring the water with various dyes only rarely stains the lining of the stone canal, the axial organ most frequently taking up the colour, the axial sinus being in communication with the top of the stone canal or madreporic ampulla. Experiments involving the blocking or extirpation of the madreporite or stone canal do not result in any impairment of tube foot function. By attaching manometers to the madreporite of *Echinus esculentus* it has been shown that there was no continous flow across it, but that changes in the internal pressure of the body cavity were reflected in the movement of liquid across the madreporite.[424] However, from experiments with eosin it would seem that this fluid came principally from the axial sinus rather than the water vascular system. More recently Fechter[233] using sensitive manometers showed that under conditions of equal hydrostatic pressure there was neither inflow nor outflow across the madreporite over a period of 24 hr. Artificial perforation of the madreporite with large holes did not affect this result. Total contraction of all the tube feet in unison resulted in the expulsion of only some 4 to 5 μl from the madreporite. Such fluid could well have come from the thin-walled stone canal due to the rise in pressure within the perivisceral coelom resulting from the expansion of the ampullae, there being non-return valves preventing the outflow of fluid from the tube foot ampulla systems themselves. Exerting a pressure on the outside of the madreporite can result in the inflation of the podia and reduction of external pressure can restrict podial extension and Fechter subscribes to the view that the madreporite acts as a pressure balancing device.

If the madreporite were indispensable for such a function, it is rather surprising to learn that after extirpation no regeneration occurs, even after quite prolonged periods of time.[192] Isolated arms of starfish will continue to move about for some time after severance from the parent animal and such arms are obviously devoid of any connection with the rest of the water vascular system. Additionally blocking of the radial water vessel of such preparations does not result in any deflation of the tube feet within a 24 hr period, although during this time the water loss from the system should have been many times that of its total volume. On these grounds alone, therefore, it seems reasonable to postulate that the topping up function of the madreporite, if it occurs at all, plays only an insignificant role and that some other mechanism must be implicated. The possible nature of such an alternative mechanism has been discussed by Binyon[62] and is based upon the chemical composition of the water vascular fluid. Although the osmotic pressure is, within certain limits, similar to both the perivisceral fluid and the surrounding medium the chemical composition is not identical. In both asteroids and echinoids increases in the potassium concentration of between 20 and 90% have been recorded. The protein content of the fluid is however, small, being about 1·4 g/l, so that the operation of a Donnan equilibrium to retain the potassium or of colloid osmotic pressure to retain fluid can be ruled out. Such a situation points to the existence of an active transport mechanism for this ion, probably located within the walls of the tube feet or ampullae. The maintenance of a hypertonicity of only 0·5 mM/l would be adequate to overcome the loss of water by ultrafiltration.

Miscellaneous Functions

The morphology and major functions of the echinoderm tube feet have been reviewed by Nichols.[579] Their range of activities include participation in respiration, feeding and locomotion, also acting in a sensory capacity. Primitively these thin-walled diverticula of the water vascular system probably functioned solely for gas exchange. The first indications of the presence of ampullae seem to occur first in animals of the Ordovician period and it is likely that such tube feet, in spite of being extensible, still lacked suckers and relied upon mucus secretion for attachment or entrapping food much as do present-day crinoids and ophiuroids. Suckers are a more recent innovation and are present in the remaining three classes of echinoderm, reaching their highest development in the echinoids especially in such forms as *Echinocardium*, where they have become diversified to meet a wide range of requirements.[575]

Sucker Attachment and Mucus Secretion

The secretion of mucus by echinoderms, especially over the general body surface, has been frequently investigated often in connection with researches unassociated with locomotion. Some 56% of the total adhesive capacity of the tube foot of *A. vulgaris* was due to suction and the remainder to other factors, of which stickiness of the mucus secretion was thought to be the most important.[609,610] It was further determined that the average suction force which could be exerted by a single podium was in the region of 29 g. When it is realized that a large starfish may possess several thousand podia, the large force necessary to remove a firmly attached animal from the substratum will be appreciated as will be the force which can be exerted upon the shells of bivalves (see Chapter 1). The force necessary to dislodge *Arbacia lixula* and *Paracentrotus lividus* from the substratum was found to be 4·5 and 8·2 Kg respectively.[496] Such force resulted in the shearing of the tube feet and in nature it would be limited by the tensile strength of the ambulacral structure. Special bundles of the longitudinal muscles of the podium are diverted to points of insertion within the confines of the sucker disc and are responsible for tensioning the roof of the suction cavity and thus lowering the pressure within it.[721] Radial muscles which were antagonistic to these, in that they were involved in the process of detachment of the sucker, were also described. The mucus secretion would seem to play a vital part in sealing the rim of the sucker to the substratum as well as its inherent adhesive properties. The nature and distribution of the podial mucocytes of several echinoderms is well documented.[93,191,259,718,721] In the case of certain ophiuroids the podial papillae contain both mucocytes as well as sensory cells. Electron microscopical studies have shown the presence within the epidermis of a microtubular system which appears to be responsible for the conveyance of secretion to the exterior.[114,122,123,124] This secretion consists of two types, a granular substance resembling the droplets in the mucus glands and "secretory packets", which have occasionally been observed upon the surface of the podium and are thought to contain an acid mucopolysaccharide. These packets have been shown to be elaborated in the cells, near to the Golgi body, which is often more highly developed in these cells. The microtubules appear to contain either these packets or the granular material. It is suggested that a mixing of these two components may be necessary for the development of adhesive properties. A microvillar system has also been recorded in the epithelial cells.[580] The ultrastructure of the packets has been more intensively studied in five species of asteroid.[323,730] These take the form of ellipsoidal packets, which are about $0{\cdot}75\mu$ long by $0{\cdot}5\mu$ in diameter, with a dense rim bounded by a unit membrane and containing hexagonal rods some 300Å in diameter. Each packet contains some 125 to 150 rods which appear to have a more dense outer layer with projections, which it is presumed assist in attaching them to each other. In two of the species investigated, *Patiria miniata* and *Pisaster giganteus*, additional particles 2000Å by 1500Å has also been described as being present in the channels and even coating the secretory packets. Stages in the formation of these packets are illustrated as well as their discharge onto the walking surface of the tube foot. Similar particles have been described in the basal disc of *Hydra piradi*, which differ only in some aspects of their ultrastructure and sites of synthesis and transport.[626] Their adhesive function in all cases is beyond doubt.

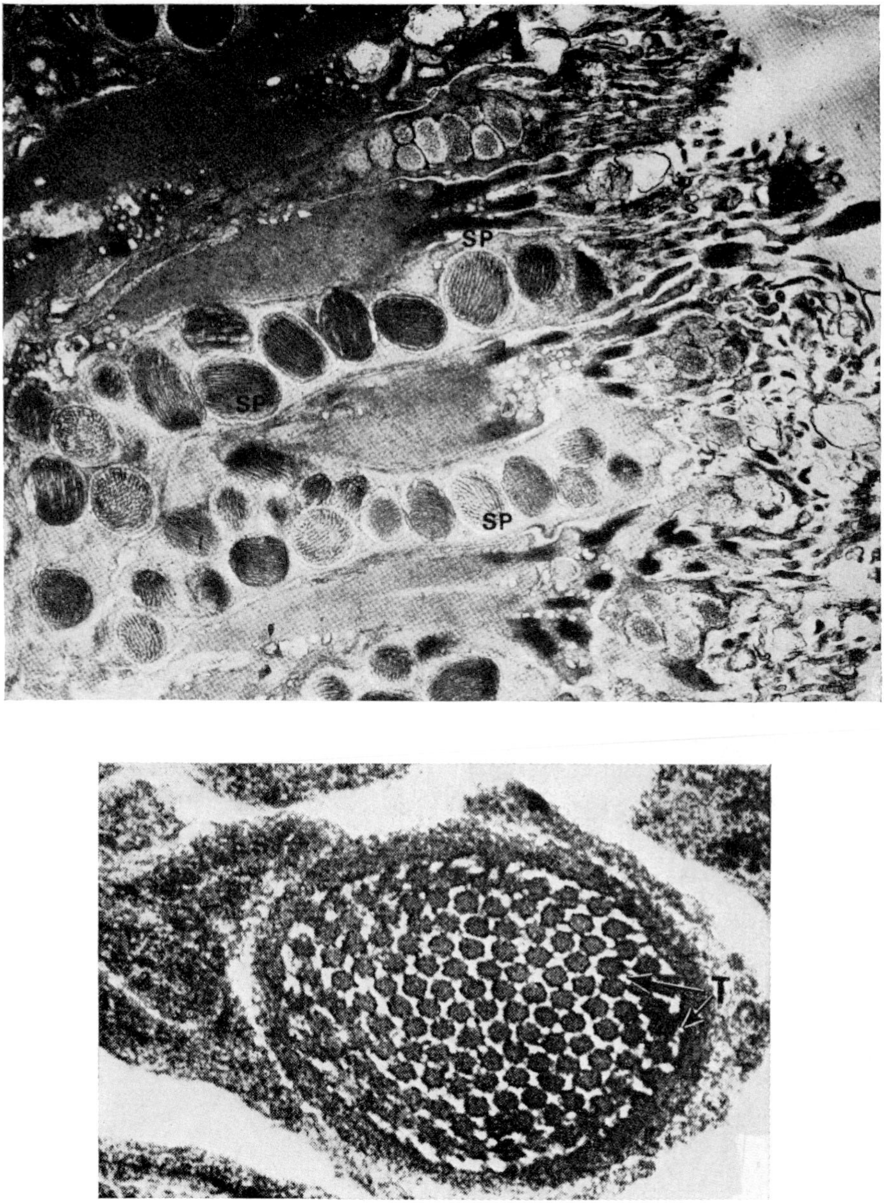

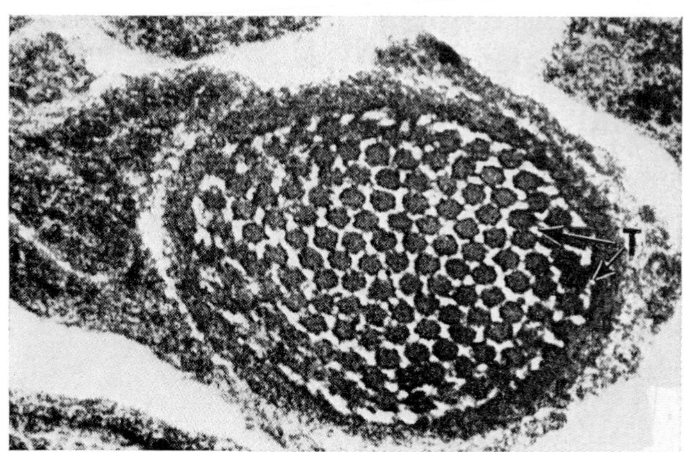

FIG. 16. Secretory surface of tube foot of *Asterias forbesi*. × 78,000. Shows secretory packets in both cross and longitudinal section (SP) in channels leading to the surface. Cross-section of a secretory packet is shown at 136,000. Note rim of material, stippled interior of rods with thread between them. (T).

(*facing page 134*)

Locomotion and Bilateral Symmetry

(i) *Asteroidea*

The exact means by which asteroids employ their tube feet in the process of locomotion has been the subject of conflicting theories which can be briefly summarized as follows: To earlier zoologists the actual attachment of the sucker was an integral phase in the stepping cycle, the ensuing contraction of postural and longitudinal muscles pulling the animal along. Such may be called the traction theory. Jennings[371] doubted this view when noting that starfish could proceed over slippery or sandy surfaces and he suggested that the podium served merely as a movable lever, thus propelling the animal along. Kerkut[399] has finally produced experimental proof that both forms of locomotion can occur, the actual mode of progression employed depending upon the nature and gradient of the substratum. If the surface is a horizontal one, the animal employs its tube feet as struts, but if a resistance is encountered or the surface be inclined or vertical then the tube feet attach themselves to the substratum and a traction mechanism results. Although it is reported that certain species of asteroid are capable of some form of swimming by thrashing movements, the tube feet are the main organs of locomotion in this class. The rate of normal ambulacral locomotion obviously varies from species to species, the fastest asteroid seems to be *Solaster papposus*, which has been observed travelling at a speed of some 2 m/min.[511] Such a speed would probably not be maintained for any length of time and the total distance that an animal moves from a given area is very small (see Chapter 10). Morphologically radially symmetrical, it comes as some surprise to learn that some species of asteroid have a decided preference to lead with a certain arm in locomotion. Some arm preference in the righting reaction has already been noticed. Preyer[640] reported such a preference in only one of the species he studied, *Astropecten pentacanthus*, and similar results have been obtained on *Asterias rubens*,[66] *A. forbesi*[151] and *Pteraster tesselatus*.[674] Long-term observation upon *A. rubens* indicated a distinct preference for stepping with arm 2 in the lead.[723] The most marked example is probably that of *Pycnopodia helianthoides* where a high degree of bilaterality was observed.[406,407] However, some workers have failed to confirm such preferences in *A. rubens* and four other species,[199] and in *Echinaster crassispina*.[161,164] It seems most likely that even with alternation of dominance of the various arms, some persistence of dominance in any one centre might account for conflicting results if the duration of the experiment was insufficient.

(ii) *Echinoidea*

The mode of function of the suckers here described for asteroids almost certainly holds true for other classes possessing them, differing perhaps in detail. In the echinoids, where the most powerful development of the tube feet occurs, the suckers are reinforced by calcareous ossicles called the frame and the rosette. These function for the attachment of muscles and prevent the deformation of the sucker and hence improve its adhesive capacity. This system has been described in detail for two species of regular echinoid.[578] Such echinoids are able to climb vertical surfaces with the aid of their tube feet and can even manoeuvre themselves whilst clinging to such a surface so as to bring the oral surface against the substratum. Such movement requires the functioning of extremely powerful tube feet. Podial progression upon horizontal surfaces is less rapid and less frequent, the maximum recorded speed being 15 cm/min.[676] The unco-ordinated movement of the spines can result in a speed ten times as fast.[777] The animal shows no radial preference in either circumstances and seems completely omnidirectional. When out of water, the lantern can be used for a rather slow, jerky form of movement which involves the protrusion of the teeth through the mouth.[289,676] As already noted, the tube feet of the spatangoid *Echinocardium* are specialized into dorsal and subanal burrow building podia, oral feeding podia, all three types being heavily papillate, together with sensory and respiratory tube feet where the suckers are absent.[575] Muscle operated mucus glands are also recorded on the surface of the burrow builders. These urchins have developed a strong degree of bilateral symmetry and are in fact unable to move in any other direction than straight ahead. Movement is entirely by means of the spines.

(iii) *Ophiuroidea and Crinoidea*

In the remaining classes, the tube feet are somewhat less important in locomotion. In the ophiuroids the spines also are employed in locomotion, but not to the exclusion of the podia, which can exhibit a wide range of activities.[813] Grasping of objects by one or more arms can result in a rapid but jerky locomotion, and no arm preference is shown. Up to 1·8 m/min have been recorded by an ophiuroid making its typical rowing movements.[676] The tube feet of ophiuroids are capable of producing copious secretions of mucus which is mainly employed in feeding and seems to be of little adhesive value in movement. The absence of a sucker, therefore, would demand a "lever" type mode of progression. Rather similar structures to the ophiuroid tube foot papillae are to be found in the crinoid *Antedon*.[577] A mixture of sensory, mucus-secreting and muscle cells results in the ejection of mucus strings during the feeding activities of the tube feet. The tube feet are arranged in groups of three, with two groups issuing from a single compartment of the extremely specialized radial water vessel already referred to. Whilst retraction will therefore be an autonomous process, protraction will be synchronous. The mucus strings so emitted are passed by the smaller members of each group to the ambulacral groove and subsequently by ciliary activity to the mouth. The function of the crinoid tube feet would seem to be wholly unconcerned with locomotion. Their commonest form of movement is an exceptionally graceful form of swimming which is brought about by the gentle raising, followed by a more vigorous downbeat, of an alternating series of arms. Clark observed *Tropiometra carinata* to attain a speed of some 5 m/min, although at this pace they soon tire.[140] Normally attached by their cirri, these structures effect only the feeblest movement. A further method of locomotion is reminiscent of certain ophiuroids. The pinnules of the leading arm attach to some object probably by mucus secretion and the arms bearing them then contract. The arms on opposite sides of the disc push against the substratum so moving the animal along in a slow crawl. Further, a running motion can be performed by the tips of the arms, the disc being held high off the ground.

(iv) *Holothuroidea*

In those forms of holothurian possessing podia, there is a marked division into the locomotory trivium usually equipped with suckers and the sensory bivium which are reduced to warty protuberances of the body wall. Although possessing calcareous spicules of a rather different nature from those found in echinoids, the disposition of the musculature suggests that mucus adhesion may be of greater importance than the operation of a suction cup in the attachment of the tube feet. The non-locomotory podia still receive a supply from the nearest water vascular canal and are often retractile. The oral tentacles however, are relatively enormous and often finely branched and rich in glandular and neurosensory cells and are supplied with large ampullae. They are capable of quite complex movements and are the main feeding organs of the class. With the aid of the tube feet, assisted in many aspidochirotes by the specialization of the body wall to form a creeping sole, these animals are able to move with some speed. The relative importance of the body wall in locomotion varies from species to species, but its most active role is probably to be seen in *Parastichopus parvimensis* which Parker described as a "giant caterpillar".[615] The waves of contraction start at the posterior end which is moved forward a short distance. The middle and finally the anterior ends follow suit, the animal progressing at some 6 cm/min. A bounding type of locomotion in the related *Astichopus multifidus* could be elicited in captivity by a number of stimuli including sudden changes of temperature, dilute or oxygen-deficient sea water.[306] This results from the passage of a similar forwardly directed peristaltic wave along the body at a similar rate, one every 65 sec. However, this species can progress much more rapidly, a maximum of 2 m/min being observed. Characteristic rolling and exploratory movements have also been observed. Analysis of the behaviour of *Opheodesoma spectabilis* showed that it used its tentacles mainly for feeding and slow locomotion but that for more rapid movements, peristalsis of the body wall is resorted to.[52] The apodous holothurians are largely burrowers, in which process the tentacles often play a part, but some progress along the surface is possible by attaching their tentacles and pulling themselves along. Some species of holothurians, on the other hand, are probably amongst the most sedentary of echinoderms, one specimen of *Cucumaria planci* being supposedly stationary for over two years whilst in an aquarium.[591]

Active swimming is rare in the holothurians. The bathypelagic elasipods, such as *Galatheathuria aspera* and *Benthodytes typica*, have a lateral brim around the body much after the manner of a cuttlefish which executes undulatory movements in swimming. However, little is known about the behaviour of most of these forms, except that many possess curious appendages thought to be associated with flotation. In preparation for natural swimming, *Leptosynapta inhaerens* inflates itself with a quantity of sea water.[158] This passes into the perivisceral coelom, not the digestive tract. Such behaviour is doubtlessly associated with buoyancy and swimming seems to occur principally during the brief young adult phase when the gut is devoid of its usual load of sand, a factor contributing further to a low specific gravity. The swimming movement consists of a curving of the body into a U-shape followed by a scissor-like flicking action. If touched whilst swimming, the animal deflates to about one-quarter of its size and sinks to the bottom. Nocturnal swimming in the adult results in the animal executing sinusoidal movements as it progresses through the water.[306] The wave, taking only about 2 sec, passes along the body from the head posteriorly, the reverse direction of the peristaltic waves in the Stichopodidae. Using a wide range of stimuli the swimming response could be elicited in captivity. Undulatory movements have been observed in another dark adapted synaptid *Labidoplax dubia.* This species proceeds anal end first and was induced to swim under experimental conditions.[356]

Although echinoderms are often thought of as a rather sluggish group of animals, many crinoids of course being sessile, representatives of every class have been shown to be capable of travelling at speeds of up to about 2 m/min. This, of course, is purely a measure of the rate they can travel when exhibiting a flight reaction. If maintained, the animal could cover some two miles per day. The fact that echinoderms will only cover short distances, even over quite long periods, will be considered further in Chapter 10.

Neurology and Co-ordination

To travel at such a speed, an asteroid for example requires the close co-ordination of several hundred tube feet, not only with regard to the stepping cycle itself, but also to the postural attitudes that these feet on different arms must adopt in order that a uniform direction of movement may be pursued. Such co-ordinating mechanisms have been the subject of many researches. The following reviews contain many references to earlier work.[399,400,719,720,723,724,725]

Asteroidea

Although the nervous system generally has been studied in some detail, consideration of it will be deferred to Chapter 11. Confining out attention solely to the tube feet, the picture which emerges is briefly as follows. The locomotor activity of the tube feet engaged in the stepping cycle is governed by the operation of two sets of antagonistic muscles. The podial retractor muscles and the ampullary muscles are responsible for the protraction and retraction phases of the cycle, the mechanical aspects of which have already been considered. They are reflex in nature and require the presence of a certain minimal length of radial nerve cord.[196] The remaining set of muscles consists of any diametrically opposed group of fibres in the circumferentially arranged postural muscles. These are inserted at the base of the podium and on the under surface of the ambulacral ossicles. On theoretical grounds it has been postulated that these must be divided into at least ten different segments, each separately innervated, so that the feet may be pointed in any required direction. The direction in which the tube feet of the leading arm are stepping is controlled by two factors, the nature of peripheral excitation and the activity of certain motor centres within the central nervous system.

Peripheral excitation is conducted to the tube foot musculature by a somewhat complex neuron chain. For the purposes of efferent innervation the aboral integument is divided along the length of the arm into two halves making what Smith terms an excitation watershed.[719] Conduction within each half is in the basiepithelial plexus. Internunciary connection is made with the hyponeural complex or Lange's nerve which is situated aboral to the radial nerve cord in the floor of the radial perihaemal sinus. Axons run from hyponeural tissue to encircle the tube feet and synapse with the lateral motor centre of which there is one on either side of the ambulacral groove in each segment of the arm. From this centre fibres run to the dorsal sheath musculature and the papulae. In encircling the tube feet the axons of the hyponeural motor system make close contact with small connective tissue pads situated within the lumena. These pads are the site of the junction with

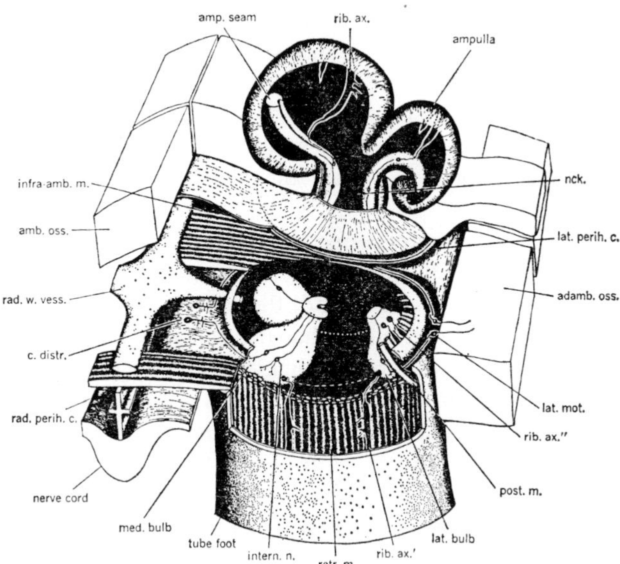

FIG. 17. Neuron systems and central nervous connections of the tube foot and ampulla of *Astropecten irregularis*. Schematic stereogram of part of an *Astropecten* arm showing some of the principal neuron systems of the intrinsic motor complex of the foot and their relations to the central distributory motor neurons extrinsic to the foot. An ambulacral ossicle and an ampulla have been removed in order to expose the cavity of the tube foot and the medial and lateral bulbs. The bulbs have an open lattice structure; they suppoıt and invest the neurons and the fibres of the intrinsic motor complex. Representative neurons included in the figure are the bulb internuncials, distributory chain neurons, and the ultimate (ribbon axon) neurons serving the retractor and postural muscles of the foot and the ampulla (foot protractor) musculature. The levels of origin of the two former series are represented respectively by the broken and the dotted lines. The latter connected to the bulb complex by chain neurons, are seen in the bisected ampulla. *Adamb. oss.*, adambulacral ossicle; *amb. oss.*, ambulacral ossicle; *amp. seam*, ampulla seam; *c. distr.*, central distributory motor neuron; *infra-amb. m*, infra-ambulacral muscle; *intern. n.*, internuncial neuron of the bulb; *lat. bulb.*, lateıal bulb; *lat. mot.*, lateral motor complex; *lat. perih. c.*, lateral perihaemal canal; *med. bulb*. one of the lobes of the medial bulb; *nck.*, ampulla neck; *post. m.*, postural (orienting) muscles of the foot; *rad. perih. c.*, radial perihaemal canal; *rad. w. vess.*, radial water vessel; *retr. m.*, retractor muscles of the foot; *rib. ax.*, *rib. ax.'*, *rib. ax.''*, ribbon axons of the ampulla, foot retractor and postural muscles respectively.

multipolar neurons by which excitation can be passed to podial and postural muscles and also of synapses with the elongate muscles tails from the ampullary muscles.[147] In addition to providing a mechanism by which external stimuli can orientate the direction of stepping of the podia, there is also evidence that peripheral stimulation is necessary to maintain the appropriate level of excitation within the nervous system generally.

Isolated arms lacking any connection with the circumoral nerve ring move, in approximately 97% of those cases examined, with the proximal region foremost. On the other hand, those arms which possess such a connection to a small section of the ring will proceed with the distal end first. Careful transection experiments so as to divide the animal into five neurologically discrete arms results in distal stepping which after a long period slowly changes to a proximal one. If the animal be similarly divided into groups of two and three arms, then there is common pointing within each group, the arm with the distally stepping tube feet being dominant. Transection of the radial nerve cord just distal to its connection with the circumoral nerve ring eliminates that arm from any form of dominance and feet distal to the cut do not co-ordinate with the remaining arms. Interradial pointing is less frequent and less stable than radial pointing and on such occasions oscillation of dominance between the two adjacent arms is to be found. Shifts of dominance in an intact animal occur in a clockwise or anti-clockwise direction to adjacent arms rather than in a random manner. If two arms on opposite sides of the animal simultaneously attain dominance, as they occasionally do in *Phataria*, then the result is autotomy across the disc.[541]

Such evidence led Smith to postulate the existence of five centres or directional pacemakers at the junctions of the five radial nerve cords with the circumoral nerve ring. To account for the co-ordination of the tube feet, nerve tracts are thought to run from these centres around the circumoral nerve ring in both directions and thence out along the radial nerve cords. Although such tracts have been observed within the radial nerve cords, there is at present no anatomical or electrophysiological evidence for the existence of such centres. The varying directions in which the tube feet can point, and Kerkut notes up to seventeen, is only a reflection of the differing levels of activity within such centres. Tensions developed within the body wall are also of some importance in determining the direction of stepping, although proprioceptive mechanisms within the tube feet are deemed unlikely.

Ophiuroidea

Investigations of podial behaviour in other classes of echinoderm have not been so detailed. In the ophiuroids the replacement of the podia by the arms as the chief organs of locomotion has resulted in a quite different pattern of motor pathways and the loss of the basi-epithelial plexus, due largely to the calcification of the epithelium, has led to changes in the sensory pathways. Because of the serial arrangement of skeletal elements, the nervous system takes on a much more segmental appearance than it does in asteroids, even to the extent of ganglion-like swellings of the radial nerve cord. As in asteroids, the tube feet receive fibres from the motor hyponeural system as well as from the sensory portions of the radial nerve cord. An experimental analysis of the behaviour of the tube feet to stimulation and their co-ordination during locomotion and other activities, similar to that made by Smith upon asteroids, is lacking. The podia show localized responses to weak stimuli and more widespread activity results from the stimulation of the integument, the more proximal the stimulus the more intense the reaction.[502] Isolated arms with no connection to the circumoral nerve ring soon become inactive, but similar arms possessing such a connection, remain active.[486,778] Such behaviour is markedly different from that shown by asteroids and indicates that the circumoral nerve ring in ophiuroids plays a much more important role, a role more like that of a conventional central nervous system.

Crinoidea

A somewhat similar function is undertaken in the crinoids, the remaining group of arm-bearing echinoderms to be considered, by the chambered organ, an organ unique to this class. It is located in the centrodorsal ossicle and is connected to the aboral brachial nerves which are of mixed motor and sensory fibres, running along each arm enclosed within the brachial ossicles. Although this system is somewhat analogous in function to the hyponeural system of the Eleutherozoa, its relationships and development rule out any homology. The similarly called hyponeural or deep oral system of the crinoids is also unique and cannot really be related to similarly called systems in other classes. By delicate work involving mechanical stimulation of the chambered organ, after severance of all connections with the ectoneural system by the removal of the visceral mass, the co-ordinating role of this structure and its associated pentagonal commissures and decussating fibres was demonstrated as long ago as 1876.[107] The maintenance of muscle tone was included in its function as a result of experiments on the swimming activity of the arms.[447] The extensive calcification of the integument, also reminiscent of the ophiuroids, has led to a restriction of the basiepithelial plexus to the soft areas such as the ambulacral grooves. Stimulation of these areas leads to localized bending of the podia or pinnules, followed by movements of the arm should the intensity of stimulation be sufficient. The superficial oral system, which also extends around the mouth in the form of a ring, is homologous with that of other echinoderms and is sensory in function. However, it only supplies the medial surface of the tube feet and its destruction does not impair conduction along the arm. The outer walls are supplied by fibres from the deep oral system which is in communication with the mixed aboral nerves. Thus, the tube feet are the meeting places of these three distinct systems. The intense activities of the crinoid tube feet as described recently by Nichols are doubtless a reflection of this extensive innervation.[577]

Isolated pairs of arms can behave as independent swimming units upon stimulation.[548] This is due to the decussating fibres, which are more distal than the pentagonal commissure, remaining intact when a more proximal transection of the arm is made.[447] As already noted, removal of the visceral mass destroys the connections between all but the aboral nervous system, and such eviscerated specimens will still swim when placed in water and can perform other functions requiring a high degree of co-ordination. There is no evidence of "active" centres at the junctions between the radial ectoneural nerve cords and the circumoral ring as in asteroids, and indeed none would appear to be necessary with such a wealth of additional nervous tissue present.

Echinoids

Making due allowance for their spherical shape, the disposition of the nervous system in echinoids follows the more familar pattern with the circumoral ectoneural nerve ring around the mouth from which the five radial nerve cords arise. The exact motor pathways from these cords to the tube feet have not been anatomically demonstrated, for there is no separation of the hyponeural tissue in this region. Presumptive motor fibres pass out in the lateral nerves to the tube feet, in which structure the basiepithelial plexus is thickened to form a distinct mesial tract. Mechanical stimulation will produce retraction of the podia, but evidence regarding the nature and role of the various portions of the nervous system in this respect is sometimes conflicting.

Several workers have shown that removal of the radial nerve cord abolishes certain responses which are initiated peripherally. Such responses include the localized retraction of the tube feet[404] and the spinal oscillations resulting from changes in light intensity.[520] It was at one time thought that the superficial input to each ambulacrum extended only as far as the middle of the adjacent interambulacrum making up the same kind of excitation watershed as was demonstrated in the asteroids.[19] More recent work has shown, however, that provided the intervening radial nerve cords have been removed, superficial excitation, presumably transmitted via some part of the basiepithelial plexus, can spread over at least three ambulacra.[533] Deeper down under the most superficial network of nerve fibres, Prouho[646a] described a more orientated stranded system extending for an unstated distance around the ambitus of the sea urchin and indeed Cuenot[179] figures small grooves on the plates in which such nerves may run. Such tracts may well be homologous with the circumferentially arranged tracts within the dorsal sheath of asteroids and could provide the anatomical basis for the results described above. Physiological evidence for separate efferent and afferent pathways with differing conduction velocities has also been presented.[533]

As with ophiuroids, there seems to be no evidence for the existence of active centres within the circumoral nerve ring, but pacemaker activity has been ascribed to the radial nerve cords. Transection of the circumoral nerve ring results in only a temporary interruption of coordination. Certain other reactions which also appear to result from some kind of central control within the animal, and are not greatly influenced by transection of the circumoral nerve ring, include the familiar righting reaction and the covering reaction of *Lytechinus variegatus* already described. Such evidence strongly suggests that contrary to the situation in asteroids, the ambulacra in echinoids have a much greater degree of autonomy, both in the control of the tube feet and other peripheral activities.

Holothuroidea

The ambulacral grooves of the Holothuroidea, like those of the Echinoidea, are closed and hence the radial nerve cords possess an epineural canal. Unlike the echinoids, however, and doubtlessly correlated with the much greater development of the circular and longitudinal muscles of the body wall, is the existence of a distinct hyponeural tract accompanying each radial nerve. There is conflicting evidence regarding the course of this system at the level of the circumoral nerve ring. Nerves from the radial nerve cords supply the medial surfaces of the tube feet, the tentacles being supplied directly from the circumoral nerve ring. Information about the neural control of these effector organs seems to be almost entirely wanting, neural control of locomotion being largely a matter of the control of the body wall musculature. A more detailed consideration of the nervous system itself will be deferred to Chapter 11.

CHAPTER 9

PHYSIOLOGY OF RESPIRATION

OXYGEN consumption in an animal like an echinoderm will be limited by the surface area available for diffusion. In some members of the group mechanisms exist for the renewal of the environmental supply of oxygen and for its transport within the animal. There also exists in the group certain specializations which if not primarily concerned with oxygen transport do in fact contribute to it.

The Respiratory Surface

The retractile dermal branchiae of the asteroids, although relatively sparse in some species, obviously increase the surface area of the animal by a considerable amount and are considered to play a dominant role in respiratory exchange.[371] Such a role may also be attributed to the water vascular system[417] and this has been demonstrated in the case of *Asterias rubens*.[509] By blocking the ambulacral grooves in increasing numbers in a given animal a reduction in the oxygen consumption to some 37% of its initial value was observed, the oxygen consumption being proportional to the tension in the medium. These animals cannot adjust to a diminishing supply of oxygen and continue respiring at a reducing rate until between 95 and 100% of the oxygen in the medium has been utilized. The podia are also thought to play some part in the gas exchange of crinoids and ophiuroids, both groups incidentally lacking dermal branchiae. However, the small size of the podia must limit their contribution. In echinoids the relatively greater development of the tube feet enables them to cope with the oxygen transport in these often nearly spherical animals where surface area is at a premium.[228,732] However, in the holothurians the more warty development of the tube feet may well limit their efficiency in this direction.

Analogous with the dermal branchiae, but opening from a different coelomic cavity, are the echinoid gills. These project around the mouth, their lumen being continuous with the lantern coelom (= peripharyngeal cavity). Cidaroids and the family Echinothuriidae possess extensions from this coelomic cavity, into the main perivisceral coelom, called Stewarts organs. Such organs may assist in the transfer of oxygen into the interior of the test. In the main, these animals lack the external gills.

Other accessory structures with a real or supposed respiratory function take a variety of forms. The Porcellanasteridae, a family of asteroids lacking dermal branchiae, possess vertical lamellate grooves called cribriform organs. These number up to fourteen per interradius in some species. A current of water is maintained over these structures which appear, at least in *Ctenodiscus crispatus*, to function in sieving out detritus.[303] However, it is difficult to believe that thin-walled structures such as these, over which a continuous flow of sea water is maintained, do not also function for gas exchange.

Making certain assumptions, Steen[732] was able to show that the extended podia of a 70 g *Strongylocentrotus droebachiensis* could, under the observed differences in oxygen tension between sea water and the coelomic fluid, permit the transfer of about 19 ml of oxygen per hour. The highest experimentally determined oxygen absorption for this species was about 2 ml/hr. Whilst the figure of 19 ml is probably maximal, even making allowances for variations in such parameters as the number of podia extended at any one time, and the thickness of the diffusion path, it is evident that the podial surface can provide an adequate route for the transfer of gas. The relatively poor development of the respiratory exchange mechanism must in part be due to the animal's inability to produce movements of the environment, gross movements of the coelomic fluid and the lack of a respiratory pigment. Movement of the ambulacral fluid during the stepping cycle of the tube feet will assist in the transport of oxygen into the ampulla where it is free to diffuse into the perivisceral coelom, although tight packing of the ampullae may reduce the surface area inside the test available for diffusion. Earlier work showed that covering the ambulacral grooves reduced the oxygen consumption to zero,[417] a finding not confirmed by Steen who was able to cover four-fifths of the podial surface before oxygen consumption was affected. The importance of the podial surface has been elegantly demonstrated by the addition of luminescent bacteria to the cleaned aboral half of a *S. purpuratus*.[229] When placed in a bowl of oxygenated water the bacteria luminesced in bands corresponding to the ambulacra. Covering the podia prevented the appearance of such bands.

Pumping Mechanisms

The remaining structures to be described all have one feature in common, namely that some sort of pumping mechanism is employed to renew the sea water bathing the respiratory surface. Examples of such pumping mechanisms are to be found in all the five classes of echinoderm. These would appear to fall into two categories, those which irrigate a specialized portion of the body surface and those responsible for moving water into and out of the body, generally through the digestive tract. In the first category would be included the respiratory chamber found in the family Pterasteridae (Asteroidea). In *Pteraster tesselatus* it consists of a supradorsal membrane supported by specially modified spines called paxillae.[674] The floor of this chamber, which of course is the aboral surface of the animal, is covered with dermal branchiae. This cavity is rhythmically inflated and deflated approximately four times per minute, the average volume change being about 65 ml per inflation. The branchiae are also said to inflate and deflate in unison with the respiratory chamber, possibly as a passive response to pressure changes. Disturbing the animal leads to convulsive deflation of the cavity. The genital bursae of the ophiuroids are invaginations of the disc lined with tall cilia. Although primarily reproductive in function, water currents have often been observed through the bursal slit. Pumping action of the disc may also assist this process, especially in the case of the burrowing amphiurids.[176,303] Although no direct evidence is available it would seem that these bursae are undoubtedly respiratory.

Water is taken into the digestive tract for purposes of respiration in a widely ranging number of species, often in association with specialized absorptive regions. In the echinoids water is passed in and out of the buccal cavity supposedly for this purpose, though no significant decrease in oxygen consumption could be observed when the gut was closed off.[229] The motive power in this case is provided by the compass apparatus whose activity is controlled by the nerve ring.[776] The movements are non-rhythmic and would seem to be initiated by oxygen deficiency and stimulated by falling pH. The crinoids and holothurians both take water in through the anus. The raised anal tube of crinoids has been observed to undergo rhythmic pulsations, but whether its purpose is respiratory or sanitary is unknown. In the holothurians the posterior portion of the intestine is dilated to form a muscular cloaca, the anus too being provided with a powerful sphincter muscle. In the Dendrochirota, Aspidochirota and Molpadonia two asymmetrically developed respiratory trees open from this cloaca. Although ramifying the viscera, these structures are completely water-tight and are, supposedly, impermeable to chloride, sulphate, sugar and urea.[338,342] The normal cycle of events begins with the opening of the anus. The cloaca then dilates, due to contraction of the suspensory muscles which are attached to the body wall, thus drawing water in. Closure of the anus and contraction of the cloaca drives water into the trees, the entrance to the intestine being closed up to this point. Water is expelled by the relaxation of the anal sphincter, followed by contraction of the cloaca, body wall and the trees themselves, even the terminal vesicles of which contract.[96,338] The passage of oxygen across the walls of the respiratory trees was elegantly demonstrated by placing an oxygen-deficient *Holothuria tubulosa* in well-oxygenated water after reduced methylene blue has been injected into the coelom.[53] Upon opening the animal after 2 hr, the coelomic epithelium was coloured blue and so, significantly, was the retia.

This cycle of events, called cloacal pumping, is subject to various forms of rhythmic activity. In *Thyone briareus* each inspiration is followed by an expiration,[616] but in *H. surinamensis* from 10 to 16 inspirations occur before one large expiration empties the system.[169] A similar cycle also occurs in *Caudina chilensis*.[596,751] Whilst *H. grisea* was found to be generally similar to *Thyone*,[613] more recent work[574] had indicated the existence of a variety of pumping rhythms within a given species. *H. forskali* was kept in a perspex box divided in such a way that the water around the general body surface and that entering and issuing from the anus could be kept separate and if necessary have different oxygen tensions. Three distinct rhythms could be identified.

With both compartments well aerated, a large volume of water is taken in slowly and then gradually expelled, about 20 ml being involved. As the oxygen tension of the water surrounding the anus falls, the pattern of fluid exchange alters. At first, water is taken in by the normal cloacal cycle and there then follows a series of rapid and small tidal inspirations and expirations. When the oxygen tension in this compartment reaches 60% of air saturation, pumping ceases, regardless of the oxygen tension of the water surrounding the general body surface. If the oxygen tension of the water surrounding the general body surface is low whilst that being pumped is high, then a third and perhaps a more familiar rhythm becomes apparent. In this cycle a series of successive small inspirations is followed by a single large expiration. This cycle is in fact identical to that described for *H. grisea*

It would seem that the stimulus for cloacal pumping originates in the posterior region of the body wall adjacent to the cloaca and requires the integrity of the radial nerve cord.[96,169,750] Such observations would seem to extend the range of activities of the postulated pacemaker of the radial nerve cords (see Chapter 7). Connection to the circumoral nerve ring does not appear to be essential. However, as isolated posterior portions of the animals pump continuously, it would seem that normal inhibitory phases arise at the more anterior levels of the body. Excitation reaches the cloacal wall by means of the suspensory muscles, and contractions appear to be more frequent in smaller animals.

Excised portions of the cloacal complex itself have been the subject of a number of physiological studies on a range of animals including *H. surinamensis*,[169] *C. chilensis*,[751] *Stichopus moebii*,[479] *Cucumaria frondosa*,[814] *T. briareus*[96] and *C. elongata*.[805] The rate of pulsation was decreased by urea, carbon dioxide, acids, potassium cyanide and hypotonic and hypertonic sea water. Increase in temperature increases the frequency of contractions with a Q_{10} of about 2·4 up to a maximum of 26°C. Above 32°C and below 12°C contractions cease in *H. surinamensis*, and there is an optimal range for cloacal pumping in *Caudina chilensis*. Magnesium is necessary for normal diastolic relaxation and calcium for systole. Wells demonstrated the existence of the "potassium paradox" in such preparations. This is a period of reduced muscle tone and inhibition in a rhythmic preparation following its transfer from a potassium-free solution to one of low potassium content. High sodium/potassium ratios in the bathing medium resulted in contracture and complete loss of any rhythmic activity. It is obvious, therefore, that there is an optimal ionic balance for the maintenance of pumping activity in the holothurian cloaca as there is in other rhythmic structures. Reduction in the oxygen tension results in periodic bursts of activity with augmented amplitude, with characteristics recalling those of the intact animal.[574] Lutz draws attention to similar respiratory behaviour in man at high altitudes.[479] Adrenalin causes a decrease in tone, amplitude and rate of pulsation. After prior treatment with cocaine, ergotoxin or a small ineffective dose of atropine, there is evidence that adrenalin has a stimulatory effect. This dual effect of adrenalin has also been observed in *Myxine* heart previously treated with dihydroxyergotamine.[227] Physostigmine, nicotine and histamine also decrease the amplitude and rate but increase the tone. The first two of these drugs may even initiate contractions in a quiescent preparation. Curare, strychnine and atropine progressively decrease the amplitude until all activity ceases, but pilocarpine is without effect. This subject will be referred to again in Chapter 11.

The relative importance of this cloacal ventilation to the animal as a whole has been demonstrated by covering the anus, whereupon the oxygen consumption was reduced by 50 to 60%.[811] Covering of the tentacles was without effect, so it may be concluded that this specialized portion of the water vascular system has only a very minor role in oxygen transport. It should be noted, however that eviscerated specimens, lacking respiratory trees and cloacal complex, will regenerate in well-oxygenated water suggesting that the general body surface must at least be adequate for this purpose. H. L. Clark in 1899 described slit-like canals linking the rectum with the coelomic cavity in *Synapta vivipara.*[139] He concluded that these canals functioned for the inward transport of sperm and for the egress of young. Fluid movements were also thought to occur via this route. Similar canals linking the perivisceral coelom with the anus in *C. chilensis* have been termed coeloanal canals.[390] These canals are 40 to 50μ in diameter and a reddish coloured coelomic fluid was seen to issue from them on occasions. Apart from a further reference by Tao,[751] their existence has been overlooked in recent years, although water movements into and out of the coelom have been observed in several species. *Leptosynapta*, for example, can contract to about one-quarter of its normal size and reinflate, but the source of the fluid is unknown, the intestine remaining a similar size throughout.[158] Whether these canals play any part in fluid exchange in the pumping cycle of normal or eviscerated animals is unknown. In a recent paper, also upon *Leptosynapta*, fluid was observed issuing from five pores around the anus, situated at the termination of the longitudinal muscle bands.[17] It was confirmed that these specimens could also contract to about one-quarter of their normal size. The source of the fluid for reinflation is also unknown.

Coelomic Transport

Once oxygen has diffused across the surface of the animal's body, or specialized respiratory structure, it has to be transported to the various organs and tissues. Movements of the body wall during locomotion in such animals as holothurians and the arms of asteroids and ophiuroids will in all probability permit quite large redistributions of the coelomic fluid. The coelomic cavity is also lined with ciliated epithelium and this too contributes to fluid movements.

The rapid elimination of indian-ink-laden corpuscles through the dermal branchiae indicate that such transport is quite significant.[97,212a,291,367] The rocking movements of Aristotle's Lantern during either feeding or locomotion must also contribute to the circulation of coelomic fluid in echinoids where a rigid test must reduce circulation due to other causes.

Respiratory Pigments

The presence of a respiratory pigment in the coelomic fluid would obviously increase the efficiency of oxygen uptake. Amongst the variety of corpuscles to be found floating in the coelomic fluid of echinoderms are variously coloured, often motile, cells called haemocytes. Although principally found in holothurians, some are to be found in the water vascular system of ophiuroids such as *Ophiactis virens* in what must be one of the earliest records of haemoglobin in echinoderms.[253] Other instances of the characterization of haemoglobin in echinoderms include *Thyonella gemmata*,[357] *Thyone briareus* (Polian vesicle),[783] *Cucumaria frauenfeldi*,[351] *Caudina chilensis* and *Molpadia roretzii* (haemal vessels, polian vesicles, tentacular ampullae and perivisceral coelom),[409] *Cucumaria miniata*,[167] *M. arenicola*[491] and histochemically in *C. lubrica*, *C. piperata*, *Eupentacta pseudoquinquesemita*, *E. quinquesemita* and *Pentamera populifera*.[345] The absorption spectra of these pigments are in many cases very close to those of mammalian haemoglobins. The molecular weight of that of *T. briareus* has been determined by means of the ultracentrifuge and was found to be in the region of 23,600.[746] Such a low value suggests that it is nearer in size to the single unit haemoglobin of the cyclostomes, a point already referred to in Chapter 5. Oxygen dissociation curves have been published for *Caudina chilensis*,[391,751] and for erythrocyte suspensions from *Cucumaria miniata*[491] yielding $T_{\frac{1}{2}\ \mathrm{sat}}$ figures of 8·0 and 12·5 mm Hg respectively. The oxygen-carrying capacity for the coelomic fluid of *Caudina chilensis* is given as 6·1 volumes per 100 ml of fluid by Kawamoto[391] and between 9·4 and 12·4 volumes per 100 ml of fluid by Tao.[751] The extent to which these pigments function as true respiratory pigments is still obscure, and oxygen storage would seem a more likely role. As long ago as 1928 it was noted that the coelomic fluid of *Cucumaria frauenfeldi* was port wine coloured when freshly drawn, but assumed a scarlet hue on shaking with air.[351] Although no oxygen dissociation curve was produced, it appeared that the unloading tension was not so low as to prevent its functioning as a respiratory pigment. The recent and extensive work of Manwell and his colleagues upon the structure of holothurian haemoglobins and its evolutionary significance has already been referred to (Chapter 5). The haemoglobin of *C. miniata* in common with the respiratory pigments of a number of invertebrates, does not exhibit the Bohr effect.[491] The critical tension is high—100 to 115 mm Hg—and is nearly at the level of the oxygen tension of sea water—130 to 150 mg Hg—which suggests that the animal has little ability to compensate for conditions of low oxygen tension, a point borne out by the studies upon actual oxygen consumption of the whole animal. Manwell also points out that the *n* values of less than 1 obtained by Kawamoto[390] (1926) are indicative of negative haem–haem interactions, an unusual state. His figures for *C. miniata* were between 1·3 and 1·4 and the properties of the pigment were in this and other respects similar to those for other haemoglobins. Little is known of other pigments such as molpadin, which are only doubtfully associated with respiration.[167]

Respiratory Enzymes

Enzymes associated with respiration have been detected in the perivisceral fluids of *Asterias*, *Arbacia* and *Thyone*, catalase and peroxidase being detected in extracts of the respiratory trees.[782] The presence of catalase was later confirmed by Bertolini, who was unable to detect any indophenol oxidase or peroxidase.[54] Generally only low carbonic anhydrase activities have been recorded from echinoderms, except in the oesophagus of *Heliocidaris erythrogramma*, the tube feet of *Coscinasterias calamaria* and the gonads of certain asteroids. The carbon dioxide content of the coelomic fluid is recorded as being between 5 and 6 ml per 100 ml of fluid, only slightly in excess of that for sea water (4·8).[213] Other authors quote between 6·6 and 10·8 ml in *Caudina chilensis* and further details may be found in Table 3.

Haemal System

A system classically implicated in respiratory physiology is the haemal system. Although present throughout the phylum, its development only warrants the description of a "significant organ system" in the holothurians and to a lesser extent in the echinoids. However, in many instances the walls of the system are often indistinct, and Hyman[362] prefers the term lacunar system. The fluid contained in the haemal system would appear to contain no significantly greater amounts of nutrients than is to be found in the other coelomic fluids and this aspect has already been referred to in Chapter 1. Its blood-like properties are ill-defined and in the crinoids it is said to be a highly coagulable and proteinaceous fluid, but no circulation or fluid movement has been observed. In asteroids several regions of the system are contractile, including the axial gland, its terminal process and the aboral ring with the gastric haemal tufts.[290] There is probably some movement of fluid, but this is most likely to be tidal in nature rather than a true circulation. The presence of independently motile amoebocytes within the system has undoubtedly misled observers into thinking that movement of the fluid itself was taking place. No localized contractions of the ophiuroid haemal system have been reported, nor is there any evidence of fluid circulation. Although more extensively developed in the echinoids, with in many cases a very actively contractile axial complex, no circulation of fluid within the main channels of the test or viscera has been described. Just about every probable function has been ascribed to this enigmatic structure. Millott and Vevers[535] list no less than seventeen authors who have variously considered the axial organ to be excretory, genital, vestigial, pigment-producing, circulatory, this latter suggestion being the most popular with ten adherents. Certain studies upon the axial organ have already been mentioned in this book, but some more recent investigations must now be briefly referred to.[79,80,104,525,535,536] Upon closer examination by these investigators, the axial organ has proved to be even more complex than had already been believed. The haemal sinus itself has been the subject of a number of electron microscopical studies.[204,480] The wall of the sinus is permeated by numerous spaces or poorly differentiated canals referred to by Millot as the lacunar system. Such canals provide a means of communication between the perivisceral coelom on the outside and the axial sinus on the inside, which cavity communicates further with the water vascular system. The axial organ is therefore at the point of confluence of these three coelomic systems, a point which has been overlooked in the past. Inside the axial sinus is the central contractile vessel containing haemal fluid. At the aboral end of this vessel is a two-compartment pulsating chamber, whose beat would appear to be myogenic in nature, being slowed by acetylcholine and accelerated by adrenalin. There are no valves between these compartments and fluid is shown entering the first of these from the rectal haemal vessel and also by way of ostia in the walls. A circulation of fluid through this focal point of the haemal system is described by these authors, but it is questionable whether this is a true circulation or merely a tidal ebb

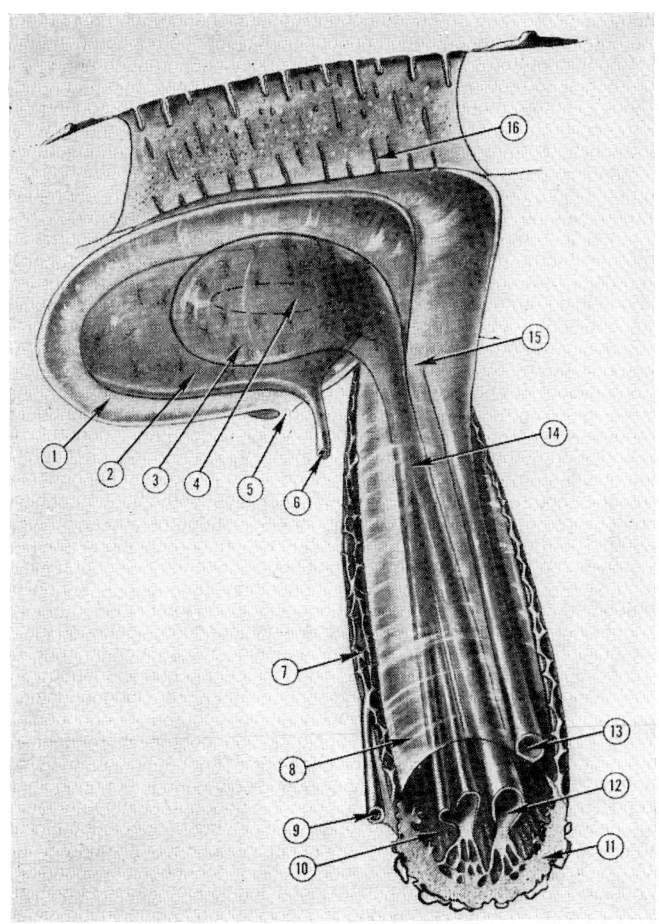

FIG. 18. Schematic longitudinal illustration of the axial gland and pulsating vessel complex of the echinoid *Strongylocentrotus purpuratus*. 1, cavity of the axocoel; 2, broad flattened compartment; 3, small conspicuous compartment; 4, opening between 2 and 3; 5, slit between axocoel and coelomic cavity; 6, rectal haemal vessel from inner haemal sinus; 7, surface vessels; 8, membrane; 9, axial oral vessel; 10, lumen of axial gland (axial sinus); 11, dense region; 12, right branch of the pulsating vessel within lumen of the axial gland; 13, stone canal; 14, single pulsating vessel near the aboral end of the axial gland and within the axial lumen; 15, opening between axial lumen and stone canal; 16, madreporite. ($\times$ 40.)

(*facing page 156*)

and flow.[104] Whilst the contractile nature of this system is not in doubt, the designation "heart" by some authors is an unfortunate one on several grounds. For example, records show that the oxygen consumption of both *Arbacia punctulata* and *Strongylocentrotus droebachiensis* is little changed by complete excision of the axial complex.[229] Verchowskaja, as long ago as 1931, found that *Asterias rubens* could live for at least six months after extirpation of the axial organ, the only noticeable effect being an enlargment of the Tiedemann's bodies.[788] She concluded from this that the axial organ was excretory. Such reasoning depends, of course, upon what function is ascribed to these latter structures. Neither implantation of supernumerary axial organs nor the excision of its own had any effect upon the animal's rate of feeding or locomotion. If this structure were a true heart, such results would surely be unlikely. The processes leading to the regeneration of the axial organ of *Arbacia punctulata* are complex.[528] Following dedifferentiation and histolysis, a blastema is formed by active proliferation and invasion by surrounding tissues including the peritoneum, stone canal lining and haemal vessels. A plate- or cup-like rudiment forms within the blastema, which upon further invasion from outside gives rise to the canalicular system and also comes to enclose the contractile vessel. The head process can reform independently from the rest of the axial organ, which seems to exert a morphogenetic influence, for where there is no regeneration of this structure, there is no regeneration of the stone canal either. Recently, however, its strategic situation and glandular nature has been emphasized, as also has its possible participation in an immune response[525] (see Chapter 6). Its histochemistry too has been examined in some detail.[536]

The Dorsal Sinus

In holothurians the radial haemal system is weakly developed, whilst that surrounding the gut and respiratory trees shows a multiplicity of vessels with anastomoses. It has been minutely described in *C. chilensis* including a directional fluid flow together with a sort of circulation, the exact pathway of which is difficult to make out.[390,751] The contractile mechanism responsible for this movement is not in this case the axial

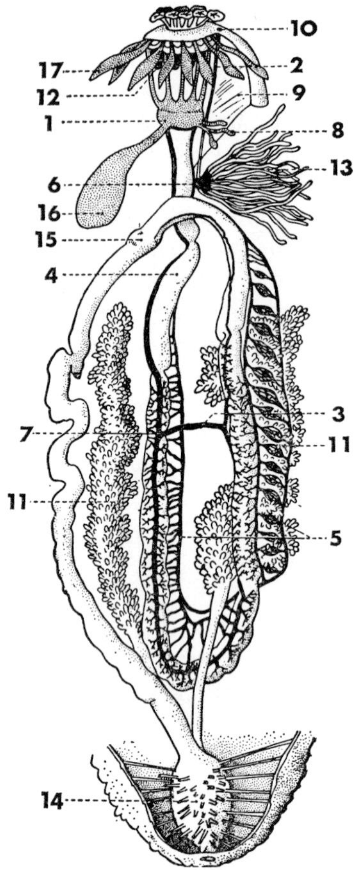

FIG. 19. Diagram of the digestive system and haemal system of *Holothuria tubulosa*. 1, circumoral water ring; 2, gonoduct; 3, transverse connection of the two portions of the gastric haemal system; 4, stomach; 5, ventral haemal sinus; 6, genital haemal strand; 7, dorsal haemal sinus; 8, madreporite and stone canals; 9, dorsal mesentery; 10, gonopore; 11, respiratory trees; 12, pores connecting coelom and peripharyngeal space; 13, gonad; 14, cloaca; 15, start of rectum; 16, Polian vesicle; 17, ampullae of tentacles.

sinus but the dorsal sinus of the intestinal vessel, and it too has been intensively studied.[220,338,390,644,814] The rate of beat has been variously observed to be between 4 and 12 per minute, a figure which compares closely with the values of between 3·5 and 10 per minute, obtained for the beating of the stomach vessels of *Echinus esculentus*.[104] However, the experimental conditions will influence the rate quite considerably, for example, the degree of distension is of greater importance, a flaccid vessel being almost motionless. Like the axial organ, the beat of this vessel is reported to be myogenic in nature. Electrical stimulation, whilst causing a localized contraction, does not affect the rate of pulsation.[338] The response to drugs may be somewhat erratic. It was found that adrenalin accelerated the beat whilst atropin slowed it.[814] A high but variable response is recorded after application of acetylcholine, inhibition being detectable down to 10^{-14}.[644] Inhibitory effects have been recorded with nicotine and tetramethyl ammonium chloride, contraction being associated with a slow wave of electronegativity.

Oxygen Consumption

Oxygen consumption figures for intact echinoderms indicate a very low level of metabolism. The figures in Table 9 have been recalculated so as to make direct comparison a little easier, although certain assumptions have had to be made in several instances. The factors affecting the level of oxygen consumption are various, but perhaps its dependence upon the level of dissolved oxygen is the most obvious. A number of workers have noted that within certain limits, the rate of oxygen consumption is strictly proportional to the oxygen concentration of the environment. This has been found in *C. chilensis*,[592] *S. purpuratus*, *Patiria miniata*[361] and *Asterias rubens*.[417,483,509] However, at both the upper and lower extremes there is some disagreement as to what occurs. Little increase in consumption was observed when the environmental level was raised above 6 ml/l[361] and the critical tension may be at a point well above that for air saturation. Incidentally it is recorded that the respiratory rate of *S. purpuratus* was higher in air than in water.[229] Conversely a linear response was also obtained between 2 and 12 ml/l.[417] Parcellation of the animal increased the oxygen consumption somewhat, presumably by decreasing the length of the diffusion path.[504] It is difficult to draw any hard and fast conclusions from experiments of this sort due to the effects of trauma. In holothurians, however,

TABLE 9. OXYGEN CONSUMPTION OF SOME ECHINODERMS

Species	ml O_2/g wet wt/hr	Reference
ASTEROIDEA		
Asterias rubens	0·04	417
A. rubens	0·024	65
A. forbesi	0·074	483
Echinaster sepositus	0·012	542
OPHIUROIDEA		
Ophioderma longicauda	0·021	542
Ophioglypha lacertosa	0·050	542
Amphiura chiajei	0·023	Wintzel quoted Hyman 362 recalculated on the basis of 1 g % of total nitrogen
Ophiopholis aculeata	0·052	
Ophiura albida	0·053	
Ophiocomina nigra	0·086	
Ophiothrix fragilis	0·113	
ECHINOIDEA		
Echinus miliaris (small)	0·077	417
(large)	0·056	417
Arbacia punctulata	0·016	81
Strongylocentrotus purpuratus	0·036	229
Echinocardium cordatum	0·034	542
Echinus microtuberculatus	0·010	
Sphaerechinus granularis	0·006	
Stylocidaris affinis	0·012	
Strongylocentrotus lividus	0·020	
HOLOTHUROIDEA		
Holothuria impatiens	0·019	542
H. tubulosa	0·021	
H. stellata	0·005	
Thyone briareus	0·013	347
Caudina chilensis	0·01 to 0·037	751
Holothuria forskali	0·00162 to 0·00465	574

Figures of Montouri (1913) are average of several widely differing values, animals of similar weight often varying by an order of magnitude.

strict dependence may not always obtain. The oxygen consumption of *T. briareus*, although uniform above 0·7 ml/l, fell rapidly below this figure.[347] Whilst the quantity of oxygen absorbed across the cloaca of *Holothuria forskali* varied from about 0·035 ml/g dry weight/hr at full air saturation down to zero at 65% air saturation when the cloacal pumping ceased,[574] that passing across the general body surface remained fairly constant at 0·016 ml/g dry weight/hr. This would suggest that after the cessation of cloacal pumping the respiratory rate would remain nearly constant, and there is evidence to show that an oxygen store in the form of the oxygen dissolved in the coelomic fluid can be used to counteract the loss of the cloacal oxygen supply and so keep the apparent rate of oxygen consumption below this value at 65% air saturation or less. Such a store would theoretically last for about 9 hr. If fully saturated water is made available to the animal after cloacal pumping has ceased, then the quantity passing across the cloaca rises to 0·057 ml/g dry weight/hr, falling back to its normal value in less than 1 hr, suggesting the replenishment of such a store. If the rate of oxygen consumption is proportional to the concentration, then the percentage extraction of oxygen from the environment should approach 100%. It has been claimed that *A. rubens* can indeed utilize the last traces of oxygen[483] but some authors quote a 95% extraction.[417] Using a flow technique, it was found that three species—*Astropecten aurantiacus*, *H. tubulosa* and *Echinocardium* sp.—could extract about half the oxygen from the incoming water.[336] Three species of *Strongylocentrotus* show a steady oxygen uptake until the water reaches about half saturation and here too it would seem that the perivisceral fluid can act as an oxygen store.[375] In air, oxygen uptake is restricted to about 18% of the maximum in water, unlike echinoids.

Salinity, pH and Temperature

Salinity, pH and temperature affect the rate of oxygen consumption. The oxygen consumption of *Asterias rubens* from the Baltic Sea (15‰) fell if the salinity was further reduced or if the pH fell below 5·5.[509] A small increase in consumption was observed as the salinity rose, to a maximum at 32‰, or if the pH rose above 7·8. On the contrary, North Sea animals (30‰) showed a rise when they were placed in a hypotonic environment.[65] It has also been recorded that both 50% and 200% sea water depresses the oxygen consumption of the same species.[483] The oxygen consumption of *S. purpuratus* was altered little by immersion in 70% and 120% sea water, but falls off outside this salinity range.[298] Expressing oxygen consumption in terms of the animal's live weight in experiments of this type may be of doubtful value, for salinity changes affect the degree of hydration and consequently the wet weights. The quantity of metabolically inert material included in such estimates has also been shown to vary with the animal's origin[422] (Chapter 4).

A regular increase in oxygen consumption of *A. rubens* occurs with increasing temperature and this appears to be maximal at 25°C.[416,509] In the echinoid *S. purpuratus* an exponential rise in oxygen consumption within the range 5°C at 23·5°C is reported.[230] When acclimatized to a lower temperature than normal, the level of oxygen consumption was raised over the whole range, an effect attributed to the formation at low temperatures of enzymes with lower activation energies. Raising the temperature, therefore, would have a diminishing effect at the upper end of the scale, which is what is observed.

Intrinsic Factors

Three closely related intrinsic factors—nutritional state, reproductive cycle and size—all influence oxygen consumption, but their relative effects are not always clear, nor easily elucidated. As a general rule large animals whilst consuming more oxygen than small ones consume less per unit weight and such a situation suggests that the efficiency of the respiratory mechanisms may influence the upper size range in such species as *A. rubens* and *Stichopus japonicus*.[229] Young animals have also been shown to consume more food than older ones and it can be concluded that the metabolism of younger animals is generally higher than in old ones.[282] The nature and distribution of food reserves has already been indicated in Chapter 1 and obviously the type of food, whether it be freshly obtained or mobilized from reserves, will influence oxygen consumption to yield its own appropriate respiratory quotient. The presence or absence of exogenous food supply has a marked effect upon the level of oxygen consumption in *Strongylocentrotus purpuratus*, which may be reduced by 50% upon starvation.[229] During the maturation of the gonads the uptake of oxygen does not increase in this urchin if it be expressed in terms of wet or dry weight. However, as the quantity of tissue has increased, the consumption figures appear to fall if related to the total nitrogen content.[299] The sum of the oxygen uptake figures for each of the various major components of the urchin was found to be roughly equal to that of the intact animal, and interestingly, the body wall accounted for over 70% of the total consumption (see Table 10). This value falls to only some 38% in large animals, which is just as well, otherwise the body wall might consume all the oxygen entering the animal.[297] The

TABLE 10. OXYGEN CONSUMPTION OF VARIOUS TISSUES OF A SINGLE SPECIES
Strongylocentrotus purpuratus

Tissue	% body weight	μl O_2/g wet wt/hr
Body wall	58	28·3
Lantern	4·8	13·8
Perivisceral fluid	27	1·1
Gut	4·6	103
Ovary	4·4	37
Testis A	6·0	145
B		62

perivisceral fluid index which increases in the larger animals is a component with a normally low oxygen consumption.[296] In the non-gravid animal the gonadial contribution was small, but became significant in the ripe animal, particularly in the case of males. This is not surprising in view of the fact that the interior of an echinoid test is well below full oxygen saturation and such parcellation procedures would remove one or other of the obstacles to a higher consumption, namely shortening the diffusion path.[732] The effect of parcellation upon the oxygen consumption of *A. rubens* has already been noted.

Some interesting oxygen uptake studies have been made on the isolated longitudinal muscle of the body wall of the holothurian *Stichopus mollis*.[286] An oxygen consumption of 24 μl O_2/g wet weight/hr was recorded, which as will be seen from Table 10 is of the same order of magnitude as those of other isolated organs, and incidentally is about one-third that of toad muscle. Little or no lactic acid was produced even after the addition of glucose. Increasing the concentration of potassium in the bathing medium to 25 mM/l, however, not only increased the oxygen consumption, which was maximal when the concentration in the medium reached 100 mM/l, but also stimulated lactic acid production, and in this instance added glucose did have an effect. Such changes in the ionic balance of the bathing medium brought about prolonged contracture with maximum tension being developed after 15 min and declining over a period of 1 hr. A deficiency of magnesium ions also resulted in high lactic acid production, but did not stimulate oxygen consumption. Altering the level of calcium did not affect either parameter. In its resting state it is suggested that this muscle metabolizes fat and in response to ionic stress switches its energy-liberating mechanism to carbohydrate oxidation, a situation somewhat similar to that of toad striated muscle.

CHAPTER 10

PHYSIOLOGY OF SPAWNING AND NEUROSECRETION

LACKING a transport system of the more orthodox type and operating at a comparatively simple level, echinoderms might reasonably be expected to lack such additional refinements as an endocrine system. Spongy tissue such as the axial organ or Tiedemann's bodies have all been suspected on various occasions as constituting such a system. It is always possible, of course, that such processes as the ripening gonads might initiate in the animal the sort of metabolic swing necessary to produce the observed patterns of food storage that occur during the reproductive cycle in certain species of starfish and sea urchins.[313] However, there is as yet no evidence of a truly endocrine organ in this phylum.

Neurosecretion

The comparative absence of a transport system does not, however, rule out the possibility of neurosecretion, where the secretory material could be transported to the target organs by the axons of the neurones responsible for its synthesis. Unger first described such neurosecretory cells in *Marthasterias glacialis* where they are located in the circumoral nerve ring and in the radial nerves.[779,780] From this nervous tissue was extracted two blue fluorescing substances with differing R_f values of 0·19 and 0·33 respectively. Upon injection, the slower moving substance elicited short-term movements followed by adhesion to the substratum. It also caused paling of isolated skin fragments. The second substance had an antagonistic effect to the first, causing long-term movements without adhesion to the substratum, together with a darkening of the skin fragments.

Three types of neurosecretory cell in the ganglia of the arms and disc of the ophiuroid *Ophiopholis aculeata* and two related species have been distinguished, but none in the podial ganglia and those surrounding

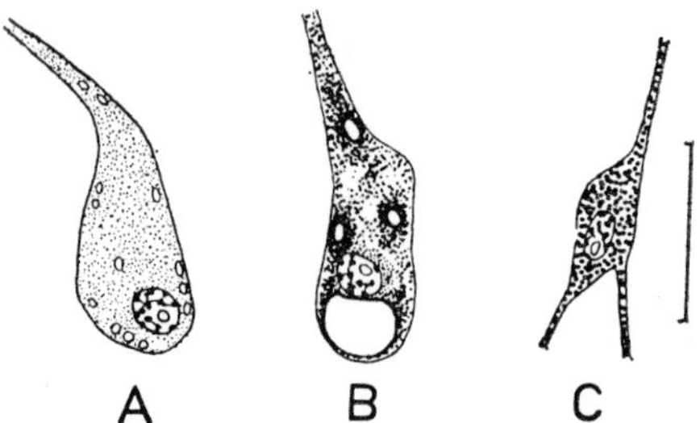

FIG. 20. Neurosecretory cells of *Ophiopholis aculeata* drawn by projection from preparation stained with paraldehyde fuchsin. Scale = 15μ.

the proximal ends of the buccal and locomotory tube feet.[258] Although the cells are often sited near coelomic spaces, there is no evidence that secretion is released through the neuronal membrane. There is, however, a diminution in the volume of secretory material along the length of the axon and axonal transport seems a reasonable suggestion, although terminal reservoirs are absent. The target organs of this system have not been identified, but instances have been cited of neurosecretory products being conveyed to regions totally devoid of musculature. The haemal system is poorly developed in ophiuroids and it is possible that such material is released directly into the coelom. Unger also described neurosecretory inclusions in the "Stutzellen" or supporting cells in the radial nerve cord of *M. glacialis*.[780] It could well be that this material could find its way via the long cellular processes into the haemal system and eventually the water vascular system. Since these inclusions contain a proportion of mucopolysaccharide, they may also form a reservoir of material for cuticle formation. This secretory system must not be confused with the neurosecretory system described below.

Spawning

The time of spawning of echinoderms is a very variable parameter and ranges from those species with only an ill-defined breeding period to those with very precise timing. Even where several species are living in close association, large variations in the time of spawning will be experienced.[294] For example, three species of *Pisaster*, all with a well-defined breeding period, exist in the same locality as *Patiria miniata*, which appears to breed throughout the year.[231] North temperate animals typically have a well-defined breeding period and cold water species have a shorter period of gonadal growth than warm water species.[74] Illumination for varying periods may affect the growth of the gonads.[77] The stimulus triggering spawning is thus not clear, although an immense amount of work has been carried out on this aspect of echinoderm reproduction.

Extrinsic Factors

In the Crinoidea, for example, the long-term observations on *Comanthus japonicus* indicate that the animal spawns annually in early October at a time corresponding with the first or last quarters of the moon and in order to maintain this timing it performs a short but precise spawn-out.[185,186] The process takes only an hour or so and commences in mid-afternoon. Isolated arms and pinnules, even under laboratory conditions, will spawn co-incidentally with those in the sea. A very specific spawning time at the end of May has been recorded in *Lamprometra kleinineri*.[555] Many holothurians have also been observed to spawn in what is probably a response to light intensity. Their spawning is sometimes associated with a characteristic elevation from their burrows followed by a waving from side to side as the gametes are emitted.[153,594] A most graphic description of this mode of behaviour has been recorded by J.-Y. Cousteau who likens their movements to those of a cobra and notes that it occurs only on a few selected days in April of each year.[160] Similar waving behaviour has also been described in the holothurian *Opheodesoma spectabilis*, together with a variety of muscular rhythms and peristalses.[52]

Although the effect of light upon echinoids has been extensively investigated (see Chapter 7, part 2) its effect on the spawning behaviour has received little attention. The echinoid *Diadema setosum* from the Red Sea spawned at monthly intervals at the full moon,[264,265,266] but specimens from a different locality did not exhibit any lunar periodicity.[556] The existence of any lunar periodicity in the north Adriatic populations of *Paracentrotus lividus* and *Sphaerechinus granularis* has been also denied.[392] Two populations of *Strongylocentrotus purpuratus* separated by 11° of latitude, where light and water temperature would have surely differed, exhibited no appreciable difference in spawning periodicity.[83,300]

Other factors which would have been considered instrumental in initiating spawning include a change of temperature. *Thyone briareus*, for example, could be stimulated to spawn by raising the temperature by 4°C and delayed by cooling.[153] Spawning in *Tripneustes esculentus* seems also to be triggered by a rise in temperature[549] as it is in certain ophiuroids.[556,605] The spawning of the deep sea echinoid *Allocentrotus fragilis* taken at 90 fathoms in the Monterey Canyon area off the west coast of the U.S.A. is thought to be initiated by an upwelling of deeper water which occurs annually.[83] Whilst the exact nature of the stimulus is unknown, the consequence of such a timing is that larval development occurs during the ensuing planktonic bloom. A further such example, this time from Antarctic waters, is the asteroid *Odontaster validus* taken in latitudes 67° and 77°.[617,618] These exhibit a reproductive synchrony over much or all of their circumpolar distribution. Spawning is timed to produce bipinnaria larvae able to take advantage of the following summer phytoplankton bloom, development being very slow. Differences in light and temperature appear to be without effect and would be ruled out by latitude anyway. Some interesting transplantation experiments have been carried out[690] using *Arbacia punctulata*, which were moved from Woods Hole, Massachusetts to Beaufort, Carolina, a change in latitude of some 10°. The animals acclimatized to their new surroundings and shifted their breeding cycle forward so as to coincide with the indigenous population. *Aequipecten irradians*, however, did not do so.[690] Two closely associated species of echinoid, *Echinarachnius parma* and *S. droebachiensis*, normally spawn in November and March respectively.[148] Hybrids of these species have been obtained under laboratory conditions. Thus it is interesting to see how different timing of the breeding cycle can isolate these two species. In Scandinavian waters *P. lividus* has differentiated into two forms, the summer and winter forms, as a result of changes in the temperature tolerance range and cleavage rates.[292]

In terms of constancy of the environment, tropical waters show some similarities with deep waters and it is true to say that in general most tropical echinoderms have an ill-defined breeding season. Some gravid individuals of the tropical echinoid *Stomopneustes variolaris* are present in the population throughout the year. Nevertheless, there is some variation in gonadial development.[301] In this instance seasonal variations in salinity could be excluded as the causal factor in spawning. Similar but more marked seasonal changes have been demonstrated in another Indian echinoderm, *Oreaster hedemanni*.[652]

Intrinsic Factors

Thus in general extrinsic factors associated with decreasing latitude lead to a prolongation of the breeding season and even to an increased number of broods, whereas animals from the more temperate zones almost invariably have only one such season. Intrinsic factors, however, also have a part to play. Nutritional factors are one such example, for the interdependence of maturation of the gonads and the food supply is a common one in animal physiology. Lipid reserves in the pyloric caeca of the starfish *Pisaster ochraceous* were high before maturation and were depleted during gonadal growth.[313] If starvation conditions prevented the animal from accumulating such reserves, then it failed to breed. Incidentally, the gonads of poorly fed populations of *Asterias rubens* are less liable to parasitization by the ciliate *Orchitophrya stellarum*.[790] In the echinoids, which lack gut caeca, the importance of the body wall as a storage organ has been indicated in several species including *Strongylocentrotus intermedius*,[280] *Allocentrotus fragilis*[295] and *S. purpuratus*.[232,454] Some material is also stored in the test and gonad itself. As long ago as 1931 Stott showed that there was a decrease in the glycogen content of the gonads just before spawning and thought that this might suggest its utilization as an energy reserve.[739] Changes in the nucleic acid component have also been observed during maturation. In male *S. purpuratus* the DNA content rises 31-fold and in the female only 5-fold. Conversely, the RNA in the male rises only 7-fold and in the female 27-fold.[313] Protein and lipids also increase, more so in the female than in the male. Such profound changes in the metabolism of the animal from a predominantly synthetic one to one intimately involved with the mobilization of such reserves and their translocation to other organs, is reminiscent of those changes found in higher animals resulting from the interplay of extrinsic and hormonal factors. It is interesting to note that steroids closely allied to classical sex hormones, which themselves have profound metabolic activity, have been extracted from the gonads of several echinoderms (see Chapter 5).

Migration

Such hormones are well known for their marked effect upon the behaviour of animals, and at the breeding season echinoderms also exhibit certain behavioural changes, although as yet these cannot be ascribed to any endocrinological phenomena. This behaviour includes pre-spawning aggregations, often the result of extensive migrations, and brooding. Migrations would seem to occur for two main reasons, food and reproduction. Few *Asterias rubens* are claimed to be present on well-worked oyster beds compared to a short distance away, but no changes in this situation were observed during the close season.[731] However, this starfish has a food preference for other molluscs and barnacles, so perhaps this is not surprising.[318] On the contrary the related *A. vulgaris* would move up to 12 m in 2 days in search of food.[716] *A. forbesi* too seems to be capable of fairly extensive movements for the same purpose[40,573] and a slight offshore migration over a period of 2 years has been observed in *A. rubens*.[320] An interesting observation only published in 1951, but apparently well known to local fishermen, records that *A. rubens* can, by taking in bubbles of air, presumably when uncovered by the tide, so reduce its specific gravity as to enable it to float. Whole populations were thought to move from place to place by this means in search of food. Whether this process is reversible or pathological is not clear.[419] Certainly personal observations on the same species in this country have indicated that when re-immersed in sea water after a period of desiccation these animals often have large air bubbles in their stomachs, but are quite unable to float. The movements of the populations of *P. ochraceous* and *Echinus troschelli* in

Departure Bay, British Columbia, have been related to variations in light intensity, the animals moving into deeper water at seasons of bright sunlight.[595] However, such movements as these are not directly related to the breeding cycle, but in view of what has been discovered regarding the pre-spawning nutritional condition of echinoderms, it is obviously important. Many other extrinsic factors are often cited as affecting the migration of invertebrates, including, for example, the presence or absence in the water of trace metals like copper, manganese and iron,[798] but little precise information is available.

So far, information upon migration for the specific purpose of breeding is scanty and conflicting. Many authors deny the existence of any such movements,[149,284,503] although some do note that these views are at variance with the opinions of local fishermen who are often convinced that some kind of mass movement does occur. Inshore migration of asteroids prior to breeding has, however, been recorded[789] and a similar inshore migration of *E. esculentus* has been observed at Port Erin.[739] Ripening *Tripneustes esculentus* in the Barbados also tend to aggregate before spawning.[463] In this country too *A. rubens* has several times been recorded as appearing locally in exceptionally large numbers, usually in the spring. Whether this is directly related to the breeding cycle or the proximity of a particularly attractive feeding ground is unknown. One consequence that pre-spawning aggregation could have however, is obvious. It will permit a simultaneous or near simultaneous spawning within a local population with the ensuing higher fertilization rate. The presence of one spawning animal either in an aquarium tank or in the field will often induce others to spawn. Such a phenomenon is well known[266] and in *S. lividus* it was demonstrated that the presence of males could stimulate other males and females, but that females could only stimulate males. A possible mechanism whereby this could occur might be that the sperm secretions could act through the aboral nerve ring, which sends a branch to each gonad.

Spawning

It was not until some 30 years later that the investigation of the chemical factors involved in spawning was reopened.[687] The muscle cells present in the ovarian wall of *A. rubens* could be induced to contract by the injection of potassium chloride into the coelom. Electrical stimulation was also effective and was employed by later workers to obtain supplies of gametes.[369] More important was the observation that the injection of aqueous extracts of whole animals into the coelom of a recipient animal caused it to spawn, although in this instance no contraction of the ovarian wall could be detected. In 1959 Chaet and McConnaughy, also using aqueous extracts prepared at 76°C, were also effective in inducing *A. forbesi* to spawn.[120] On making extracts of individual tissues, only the radial nerve cords were found to be effective. A diuretic action for this extract was also claimed although not enlarged upon. Unger's extract from the nerve cords of male *M. glacialis* was also claimed to possess diuretic as well as shedding activities.[780] In America, Chaet and his co-workers have produced a long series of papers in which the properties and possible nature of the active principle in these extracts have been examined[111, 118,121,125,127,326,365,504] (reviews 117 and 809). Parallel work in Japan has been led by Kanatani[379–382,593] The picture which emerges from all these researches is briefly this. Instead of using hot water extracts, the nerves were freed from the parent animals, lyophilized, dissolved in sea water to a strength of 5 mg% and then injected into recipient animals at a dose rate of 0·15 ml/g weight. After injection the animals often assumed a humped-up condition with the disc area elevated off the ground. After a delay of 30 to 40 min the animals spawned. Injection into one arm still resulted in all ten ovaries shedding. The action would appear to be direct upon the gonads and no substance is liberated from them which will reduce the lag phase in other animals. Extracts from other tissues of the body were without effect, although extracts from regions rich in nervous tissue such as the cardiac stomach were also effective. A shedding substance in the coelomic fluid of animals which had spawned naturally was also detectable. In general there seems to be no differential specificity between the sexes or species, but extracts from *Asterina pectinifera* will induce spawning in *Asterias amurensis* but not vice versa. The ovarian wall was shown to contract some 2 or 3 min prior to the release of eggs. Pre-treatment of the ovary with nerve extract in calcium-free sea water did not result in gamete release, but upon transferring this ovary to normal sea water,

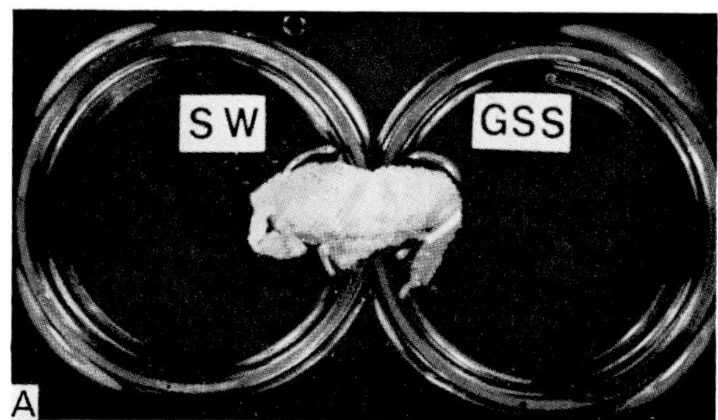

FIG. 21. Effect of local application of gamete shedding substance to isolated starfish ovary. A. An isolated ovary partially exposed to nerve extract and sea water. B. Same ovary after discharge of eggs from treated portion. Arrow indicates boundary between treated and untreated portions. SW, sea water; GSS, sea water containing nerve extract.

(*facing page 172*)

eggs were released without a second treatment with nerve extract. Eggs released by this means showed a high degree of germinal vesicle breakdown, which would suggest that the extract has several effects, namely, maturation as well as contraction of the ovarian wall. Attempts to separate these functions by fractionation have proved unsuccessful. Treatment with shedding substance also seems to reduce adhesion of the gametes both to each other and to the ovarian wall. It has been claimed recently that a meiosis inducing substance has been isolated by Sephadex gel filtration. It would appear to be distinct from the shedding substance and is very stable towards heat and trypsin digestion which will completely destroy the shedding substance; it is not thought to be a polypeptide. The shedding substance is, however, a heat labile dialysable polypeptide containing some 10 to 15 amino acids whose activity is lost after storage for 24 hr at room temperature. Gel filtration techniques upon these extracts have indicated the presence of two distinct peaks of activity and the molecular weight is probably about 5000. Attempts to mimic shedding activity with various substances with a similar physiological activity in higher animals, such as FSH have failed. Surprisingly, treatment of recipient animals with high concentrations of extract failed to induce shedding. The presence of an inhibitor was postulated, and was later named "shedhibin". Whereas the concentration of shedding substance appears to remain constant throughout the year, the concentration of shedhibin appears to vary. A fall in the concentration of the inhibitor just prior to spawning could thus act as the triggering mechanism, but what causes a fall in shedhibin is unknown.

Work upon the shedding mechanism in echinoids has received little attention. Dense suspensions of homogenized ripe sperm caused shedding in a much higher proportion of females than males in a natural population of sea urchins in the Adriatic sea.[393] Conversely, suspensions of ripe ovaries had a highly selective action upon the shedding of males. The great merit of these experiments is that they were performed upon a population of animals *in situ*. This sex specific factor has not been found in asteroids. Various drugs such as tubocurarine, hexamethonium and acetylcholine will partially inhibit spawning in sea urchins. The chemical nature of the spawning inhibitor in asteroids has recently been investigated in extracts of the testis of *Asterina pectenifera.*[364] Separation and characterization of the inhibitor by infrared and nuclear magnetic resonance spectroscopy strongly suggests that it was L-glutamic acid. No other amino acid showed the slightest trace of inhibitory activity other than aspartic acid. Bewteen 5 and 50 μg/ml of L-glutamic acid effectively inhibited the activity of 20 μg of lyophilized nerve extract, whereas 4·4 to 44 μg/ml of the natural inhibitor was required. Considering the ration of gonad to nerve, only a small part of the naturally occurring L-glutamic acid would be needed for inhibition. The mechanism, however, remains unknown, as does the nature of the inhibitor "shedhibin" from the radial nerve cords themselves.

Histological investigations have indicated the presence of granules, which in all probability are neurosecretory, in the outer regions of the radial nerve cords. No inwards migration of these particles has been observed and it is suggested that they are released into the sea. Furthermore, it is also suggested that an animal may absorb such secretion via the tube feet. Tube feet have been shown to absorb amino acids in surprisingly large amounts and pass it on into the coelomic cavity. Once there it could act upon the gonads as demonstrated *in vitro*. In the sea it could stimulate other neighbouring animals to shed, thus resulting in mass shedding, so frequently observed in aquaria and in natural conditions.

CHAPTER 11

PHYSIOLOGY OF NERVES AND MUSCLES

THE purpose of this chapter is to consider what is known of the more narrowly physiological properties of the echinoderm nervous system and musculature. These two systems are considered together, as much information has been obtained by reference to complex nerve/muscle systems. The disposition of the nervous system and the part played by it in the various behavioural patterns, such as the stepping of the tube feet, the movement of the spines and so on, has been dealt with in other parts of this book and will receive no more than a passing reference here. The criteria of nervous conduction in other animals has been well worked out and it will be of interest to see just how far such criteria apply to the nervous system of echinoderms.

Electrical Activity of the Nervous System

The most obvious specializations of a nerve are those which permit it to conduct impulses of an electrical nature along its length. Although in the past some doubt has been cast upon this aspect of echinoderm nerves, such conduction has now been amply demonstrated.[62a,532,686,748] The size of the axons in these nervous systems has been shown in recent work with the electron microscope to be extremely small.[145,388] In the hyponeural nerves entering the proximal region of the lantern retractor muscle of *Echinus esculentus* some 15% of the fibres are less than 0·25 μ in diameter and of the remainder, the largest is only 0·8 μ in diameter. It is thus of interest to explore as far as possible what form conduction takes in a nervous system with these dimensions and what similarities, both electrical and chemical, it bears to the nervous system of higher animals.

Photosensitive Nerves

The first to record electrical activity from the nervous system of an echinoderm was the Japanese worker Takahashi using the isolated radial nerve cord of the echinoid *Diadema setosum*.[748] The nerve cord was stimulated by means of a bright light and activity recorded extracellularly with silver wire. With short periods of illumination brief "on" and "off" discharges were observed with little or no intervening activity. Such a pattern of activity is reminiscent of the spine-waving behaviour described by Millott. However, the time scale for electrical activity is far shorter than that for spinal activity. Photosensitive nerves are not unknown in other invertebrates and neurons whose activity is either initiated or modified by light have been described in both molluscs[21,395] and crayfish.[396] On the contrary, no record of any electrical activity in sea urchin radial nerve cord resulting from either photic or mechanical stimulation of the spines or tube feet could be obtained from three species of echinoid.[686] Spontaneous activity arising from the ampullae during their contraction were almost certainly muscle potentials.

Electrical Stimulation and Conduction Velocity

Electrical stimulation of the radial nerve cord did, however, elicit potentials of a complex nature with a conduction velocity of between 14 and 20 cm/sec. The duration of the potential was about 200 msec and could not be recorded at distances greater than 5 to 6 cm from the point of stimulus. The magnitude had fallen to half the initial value in 7 mm. A relative refractory period follows each wave for a period of some 400 msec. There appeared to be no difference in conduction velocities for potentials travelling in an oral/aboral and an aboral/oral direction. Lengthwise division of the cord sometimes increased the magnitude of the potential and conduction appeared to be limited to certain tracts visible in unfixed preparations. Stimulation of a side branch of the radial nerve cord evoked recordable potentials only from ipsilateral branches. A single stimulus may result in up to 4 or 5 contractions of an ampulla, the ones nearest the stimulus reacting first. Facilitation can occur as a result of a burst of subthreshold stimuli

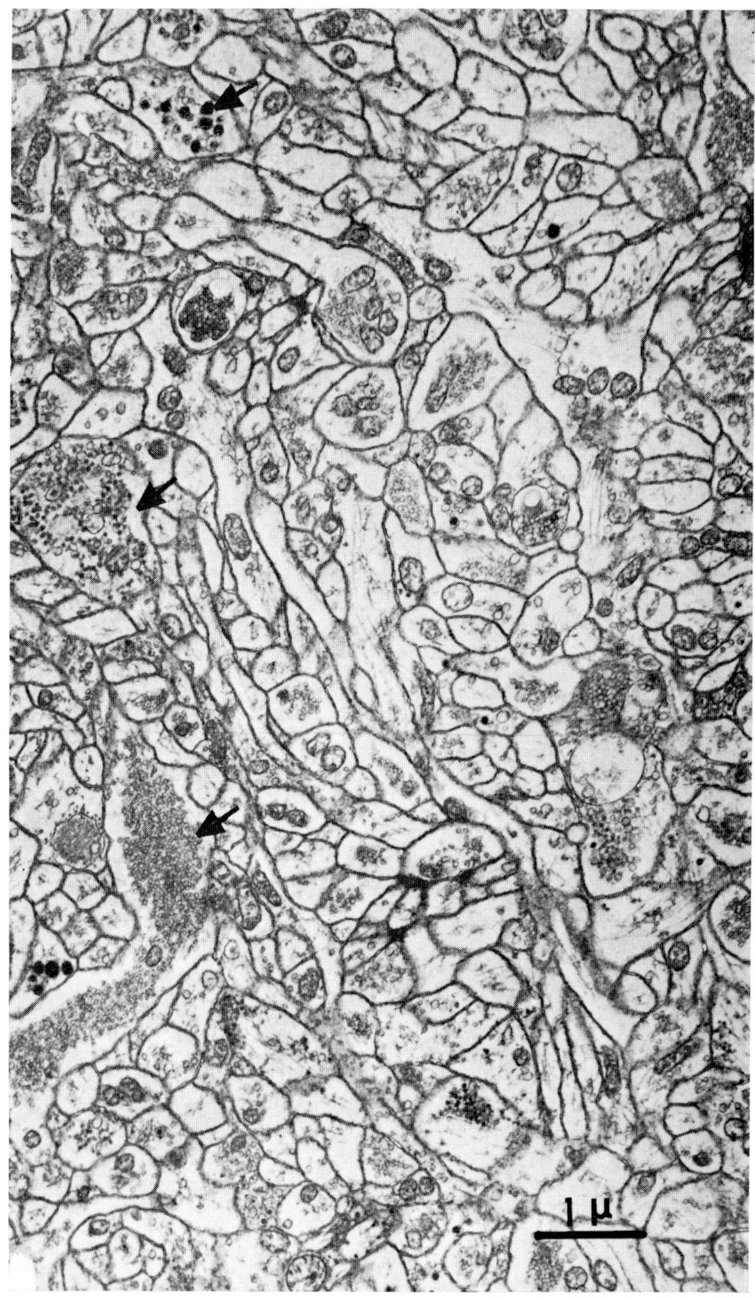

FIG. 22 A. (For caption, see facing p. 177.)

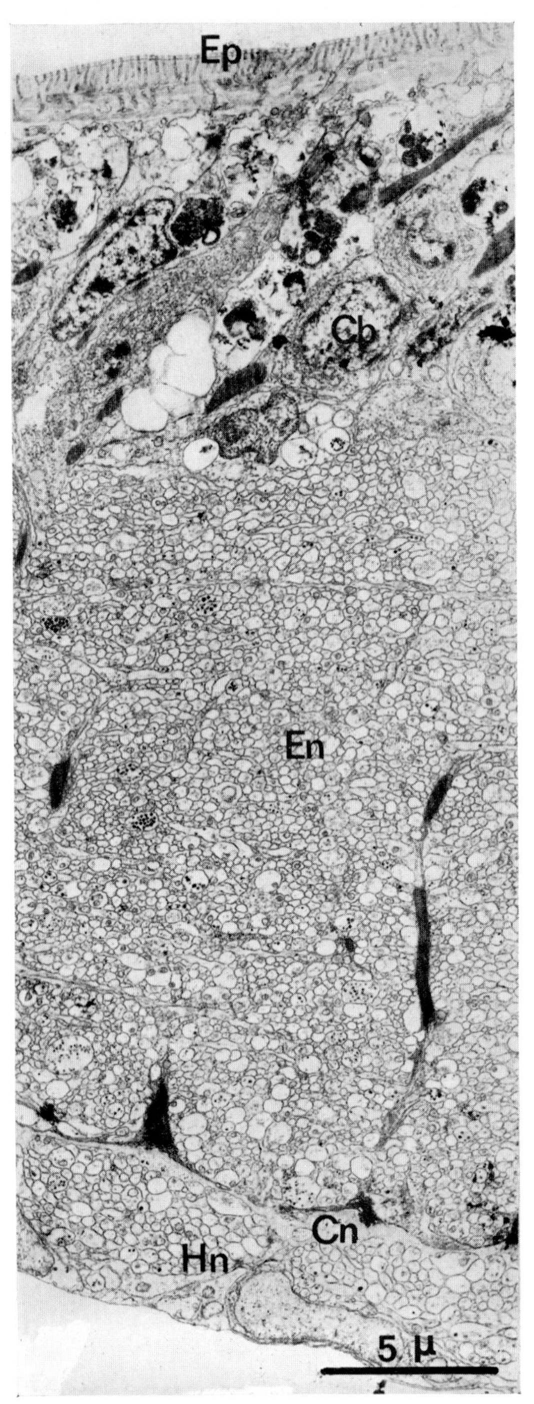

FIG. 22 B. (For caption, see facing p. 177.)

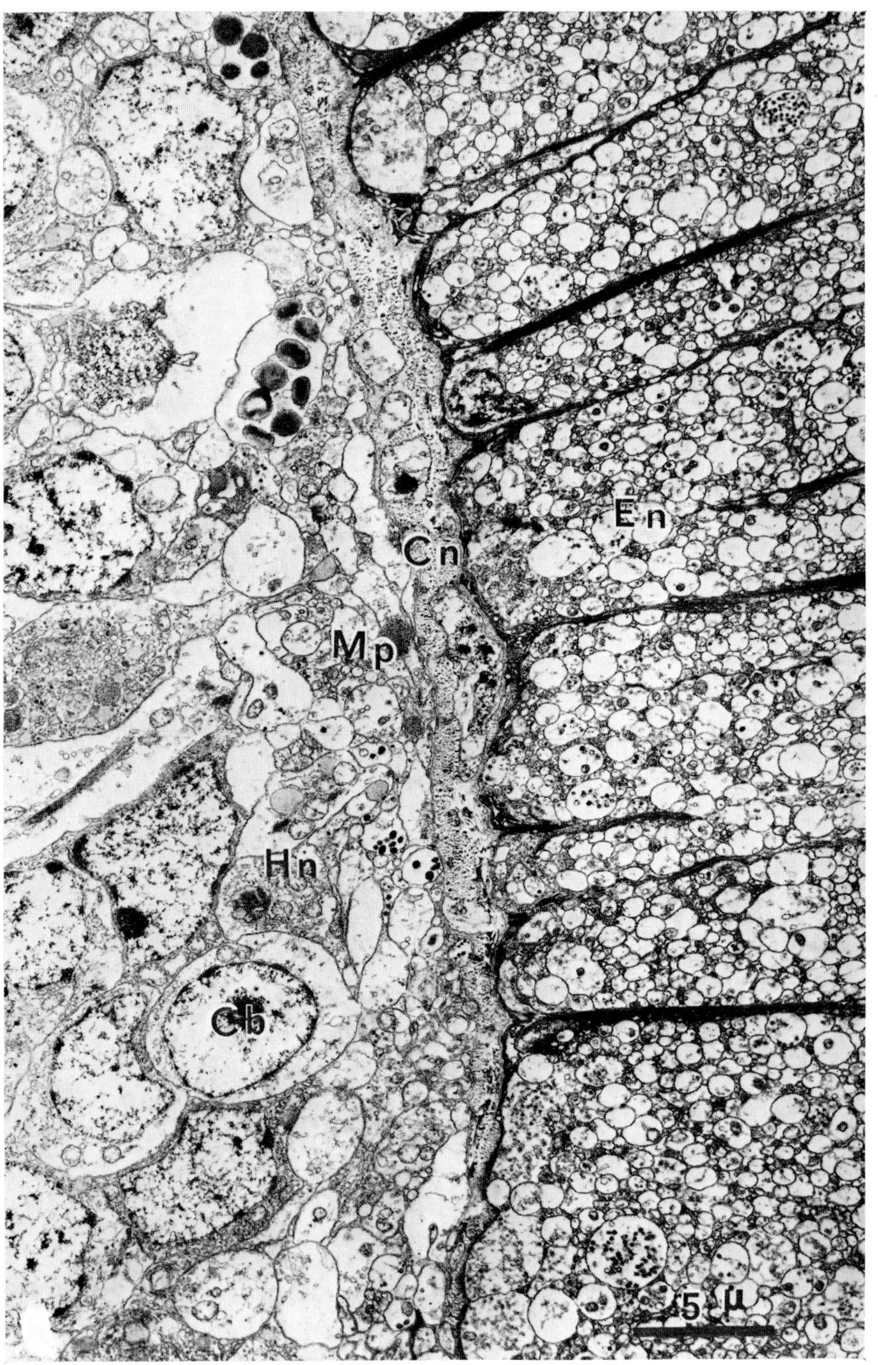

FIG. 22 C. (For caption, see facing p. 177.)

Fig. 22

A. Area of neuropile in the radial nerve cord of an echinoid showing interweaving among the constituent axons. These contain a variety of vesicles of the large and small granular and agranular types (arrows).

B. Transverse section through the radial nerve cord of a starfish cut well lateral to the midline. The external surface is covered by microvilli bearing epithelial cells (*Ep*). Cell bodies of neurones (*Cb*) occur beneath the epithelial cells and overlie the main nerve trunk of the ectoneural tissue (*En*). The dark strands are supporting fibres. The hyponeural nerve trunk (*Hn*) is separated by a narrow band of connective tissue (*Cn*).

C. Transverse section through the junctional region of the ectoneural tissue (*En*) and the hyponeural tissue (*Hn*) of the radial nerve cord of an asteroid. The layers are separated by collagenous connective tissue (*Cn*). Note the cell bodies (*Cb*) of the hyponeural nerves and a small number of processes from muscle cells (*Mp*).

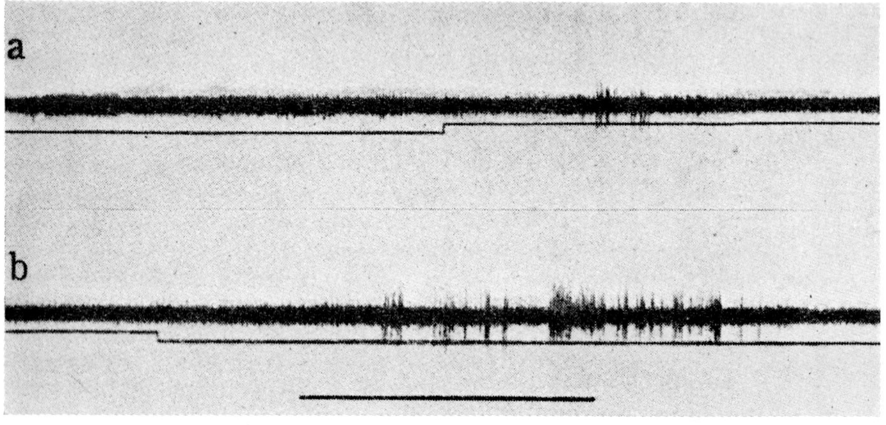

Fig. 23. Responses to light of an isolated echinoid radial nerve cord. a, "on" discharge; b, "off" discharge. Time calibration 10 secs. Temperature, 28°C.

and lasts for approximately 5 sec and spreads beyond the region of recordable potentials. Recordable potentials following electrical stimulation of *Arbacia lixula*, *E. esculentus* and *Paracentrotus lividus* are of a more simple type, consisting of a depolarization followed by a hyperpolarizing phase, the whole potential lasting about 140 to 200 msec.[532] In *D. antillarum* stimuli of 2 volts produced only a short response, but at 4 volts the more usual diphasic response appeared. Considerable decrement occurred during conduction and the period of absolute refractoriness was followed by a period of depressed excitability of a similar duration to that found in other echinoids. At higher frequencies the response reappears in an augmented form due to latent addition. The two responses in *Diadema* behaved differently to repetitive stimulation, the second and normally larger response gradually diminishing at frequencies above 6 per sec. The first response persists up to frequencies of 40 per sec. In this nerve cord there was no evidence of tracts, either visibly or by transection experiments, but evidence that conduction in the oral/aboral and aboral/oral directions is mediated through different pathways was adduced by stimulating both ends of the preparation. By timing the stimuli such that the one arrived at the recording electrodes within the refractory period of the other, no response or only a very depressed response should have been recordable. This was not so and it was postulated that different sets of fibres must have been involved. Furthermore, it was possible to link the first or fast response to mechanical movements of the spines. If a nerve which has been photically stimulated is further stimulated electrically at a level only sufficient to produce the second or slow response, during the latent period of the shadow response of the spine, or just before it, then this spinal response can be abolished, complete inhibition requiring a frequency of 4 per second. These two responses differ in their velocities of conduction, the fast one being propagated at between 18 and 33 cm/sec, the slower one at between 5 and 8 cm/sec. Nervous stimuli resulting from photic stimulation had a different velocity of conduction for the afferent and efferent impulses in the superficial nervous layer.[533] These figures were 2·2 and 6·0 cm/sec respectively and agree closely with the velocity of superficial transmission as was measured for the papula retraction response in *Patiria miniata* which was between 5 and 20 cm/sec.[99] Spinal convergence responses yielded values of between 1 and

10 cm/sec. Thus it would appear that motor responses which involve the radial nerve cord travel at slightly higher velocities than others, but there is much overlap. Impulses are conducted along the radial nerve cord of several asteroids at similar velocities to those found in echinoids.[62a] Whilst the potentials recorded were comparable to those of echinoids they were in general simpler with only the occasional trace of a second component. The absolute refractory period was about 60 msec in *Asterias rubens* with a relatively depressed state lasting for as long as 7 sec. Conduction was obviously strongly decremental and no evidence of tracts of nerve fibres conducting at higher velocities were identified.

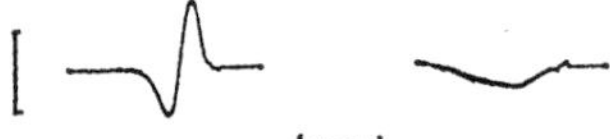

15 V, 20 msec; normal polarity. Vertical scale = 800 μV; horizontal scale = 200 msec.

reversed polarity.

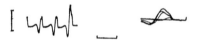

15 V, 20 msec; stimuli at 0·5 sec intervals. Vertical scale = 800 μV; horizontal scale = 400 msec.

1, 2, 5 and 10 V, 10 msec; stimuli, superimposed. Vertical scale = 400 μV; horizontal scale = 200 msec.

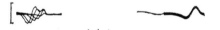

1 V, at 4, 8, 12, 16 and 20 msec; stimuli superimposed. Vertical scale = 400 μV; horizontal scale = 200 msec.

5 V, 1 msec; stimuli, 200/sec. Vertical scale = 400 μV; horizontal scale = 200 msec.

FIG. 24. Effects of electrical stimulation of the isolated radial nerve cords of *Asterias rubens*.

Due to the presence of differing thicknesses of inactive tissue in the proximity of the recording electrodes, the absolute values of the potentials recorded in these experiments have little significance. Such conditions will also affect threshold levels and should a nerve cord be moved during the course of an experiment, results will not be strictly comparable. The fast and slow response detected electrically has its counterpart in complete nerve muscle systems. Stimulation of the radial nerve cord of the holothurian *Cucumaria sykion* results in contraction of the pharyngeal retractor muscles. Such contractions exhibit a fast and a slow response, but only a single response if the muscle is stimulated directly.[633] The quick response has the lower threshold of the two, and repetitive low-frequency stimulation results in its summation, a refractory period of 100 msec following each stimulus. Different latent periods have also been established, being 0·9 sec for the fast and 3·9 sec for the slow phase. There is also a suggestion of different thresholds. In order to account for this characteristic response, the possibility of the existence of two separate conduction paths within the radial nerve cord was examined, but no such tracts were identified. It was suggested that the motor complex associated with these muscles was functioning as some kind of an amplifier. Nevertheless, the existence of separate tracts or pathways has been demonstrated in some echinoids.

Nearly every worker who has studied the nervous system of echinoderms has come to the conclusion that the radial nerve cords have some function other than that of a purely co-ordinating and transmitting system. The idea of "centres" within the circumoral nerve ring has already been referred to above. The existence of such "centres" at the junctions of the radial nerve cords and the circumoral ring in asteroids has been suggested in order to account for the changing dominance of the arms in the stepping directions of the tube feet.[723] Motor centres associated with the pharyngeal retractor muscles of the holothurian *C. sykion* have also been postulated for stimulation of this complex results in the quick response being the stronger, whereas if only the radial nerve cords were stimulated, the slow response predominates.[633,635] Two stimuli applied in succession to the radial nerve cord results in augmentation of the delayed response. This facilitation would seem to occur at the level of this centre and not at the neuromuscular junction and comparison is made to post-tetanic potentiation which occurs in the vertebrate nerve cord. This complex may thus be responsible for, or modified by, endogenous nervous activity. It can also be inhibited by the activity of antagonistic muscles. Three primary muscular activities occur in the holothurian *Opheodesoma spectabilis*. A rhythmic contraction and extension of the tentacles such as is employed in feeding and slow locomotion, peristalsis of the body wall in rapid locomotion and contraction, bending and re-extension of the anterior portion of the body wall. There exists a diurnal rhythm of activity in which transmission along the radial nerve cord plays a part and it is concluded that the nerve ring functions as a link between various active centres.[52] Much of the spontaneous activity of the spines of *D. antillarum* is thought to arise from activity within the radial nerve cords of a kind of pacemaker, which may also function in keeping up the general level of excitability in the animal.[533] A similar pacemaker would also seem to exist within the hyponeural tissue of *E. esculentus*, for spontaneous lantern movements require a certain minimal length of radial nerve cord for their continuance.[144] Severance of the nerve ring does not seem to affect the rhythms described in *Opheodesoma* and the buccal opening rhythm is apparently resistant to magnesium sulphate narcosis.[52] It should be pointed out, however, that as yet no spontaneous electrical activity has been recorded from the nervous system of these animals.

Circumoral Conduction

Conduction around the circumoral nerve ring would appear to conform to a similar pattern to that found in the radial nerve cords. Conduction is not polarized and can pass with equal facility in either direction and is decremental.[400,634] There is no evidence of frequency facilitation, nor any form of proprioceptive relay. Spontaneous slow contractions occur in the pharyngeal retractor muscles of *C. sykion* and present quite a different time course in the development of tension from the previous quick and delayed responses. These can be elicited by stimulation of the radial nerve cords.[635] Conduction of the impulses for these contractions around the circumoral nerve ring would appear to be without decrement. The response of the muscles appears to be somewhat random, those farthest away from the stimulus may contract, whilst those adjacent to the stimulus may not do so. Such contractions cease if the motor centres are destroyed or the longitudinal body wall muscles stretched. Essentially similar results were obtained from the lantern retractor muscles of *E. esculentus* when the nerve cord was stimulated mechanically,[144] facilitation being notably absent. In a passing reference to two species of echinoid, in a paper essentially denoted to the nerve net of the Actinozoa, Pantin noted that the spine response of *Strongylocentrotus lividus* and *Arbacia punctulata* was unaffected by variation in the intensity of the stimulus, but exhibited facilitation.[612]

From a consideration of conduction velocities and other parameters, it is obvious that the events taking place in the echinoderm nervous system are some two orders of magnitude slower than those recorded from the majority of animals so far investigated, and are more comparable with results obtained from the *Coelenterata*. The general slowness, both in the speed of conduction and of the duration of the action potential if that it is, leads to some doubt as to whether these electrically induced changes are in any way a reflection of nervous events which occur in the living animal. Whether these observed electrical changes, which still however follow a more or less conventional pattern, are brought about by ionic changes similar to those which have been established elsewhere, is yet to be elucidated.

Synapses

Transmission along a nerve trunk most frequently involves the participation of synapses. Areas thought to represent axo-axonic junctions have been described in electron microscopical studies upon *E. esculentus* hyponeural tissue.[145] Whether there is a fusion of the pre- and post-synaptic membranes is not yet clear, and the classical controversy of continuity versus contiguity is not finally resolved in echinoderms. There are, however, a number of vesicles present in these areas, but their function is unknown. Cholinesterase has been shown to occur in the radial nerve cord of *Asterias forbesi*,[100] but although the role of acetylcholine in the nerve cord has not been established, the situation at the neuromuscular junction will be referred to in detail later in the chapter.

Electrical Activity of Muscle

The electrical activities of echinoderm muscle have similarly received but little attention. DuBuy was one of the first to record muscle action potentials and demonstrated both slow and fast components from the pharyngeal retractor muscle of *Thyone briareus*.[209,210] After curarization, no action potentials were observable, and only very localized contractions were still visible. More distal untreated portions of the same muscle would, however, still contract even if it was the curarized portion which was stimulated. These muscles produced sustained contraction in the presence of 10^{-9} acetylcholine. These experiments point to some form of intramuscular innervation for the transmission of excitation and it is possible that the electrical activity which was observed was in part due to such nerves. The velocity of conduction of the fast component of the action potential from this muscle has been shown to be about 26 cm/sec.[643] The longitudinal body wall muscle of the same animal is innervated at frequent intervals from the radial nerve cord. Isolated preparations of this muscle exhibited a wave of electrical activity travelling at a speed which varied from 17 to 36 cm/sec and with a decrement reminiscent of echinoderm nerve.[643] Potentiation by eserine is suggestive of a cholinergic system. No potentials were recordable until the muscle, which contracts upon dissection, was stretched to its resting length. Furthermore, cutting the radial nerve cord did not abolish the response, and excitation is almost certainly transmitted through a superficial nerve plexus.[337,750]

Investigations upon the electrical properties of visceral muscle from two species show that the velocity of conduction was considerably slower than for the pharynx musculature, being 4·8 cm/sec in *T. briareus* and 5·9 cm/sec in *Arabacia punctulata*.[645] As before, stretching the muscle increased the speed of conduction, the maximal effect being observed by doubling the muscle length when the conduction velocity doubled. Conduction in these intestines is not blocked by curare.

Additionally, sensitivity to stretch, which is also characteristic of those visceral muscles where conduction is known to be on a fibre to fibre basis, suggests that echinoderms may be similar in this respect. Electron microscopy has shown that the lantern retractor muscle of *E. esculentus* is not innervated beyond the first 100 μ. Thus it is possible to record from muscle fibres without the presence of nerve fibres.[143] The muscle action potentials were shown to travel at 4 cm/sec similar to those values for visceral muscle.[645] The faster conduction of impulses in the pharyngeal retractors and longitudinal body wall musculature may possibly be due to records being taken from included nervous elements. By stimulating different regions of the hyponeural ganglion either the quick or the delayed response could be obtained from the muscle, finally indicating that the delayed response is due to neural interaction within this ganglion and not to dual innervation of the muscle.[143]

Intramuscular Conduction

Evidence as to the manner by which transmission of an impulse from one fibre to another occurs has been provided by electron microscopical studies. Membrane specializations which are supposed to have a role in such transmission between ampullary muscle fibres have been described in *Asterias rubens*.[39] Similar structures in the membrane of the podial muscle cells of *Hemicentrotus pullcherimus* and processes of the luminal epithelium ramify the muscles.[387] Such processes may serve a conduction function, for similar processes of the epithelium are associated to form the podial nerve. In a detailed study of the lantern retractor muscle of *E. esculentus*, "peg and socket" junctions between the adjoining muscle fibres exist.[146] Such junctions may be elaborated into complex interdigitations of varying forms. Mitochondria and granules, possibly glycogen, are present in such regions and although physical interactions between these cells are readily understandable, the transmission of electrical excitation is more puzzling. The muscle fibres run the entire length of the muscle, which may be up to 2 cm long and individually are quite capable of transmitting an electrical impulse.

Neuromuscular Junctions

The muscle is only innervated at its proximal end and the neuromuscular junctions all occur within the first 100 μ. These junctions are characterized by the expansion of the muscle cell into wing-like processes, enveloping the nerve termination. An aggregation of vesicles is to be found at the pre-synaptic side and many mitochondria at the post-synaptic side of the junction. Few specializations have been found

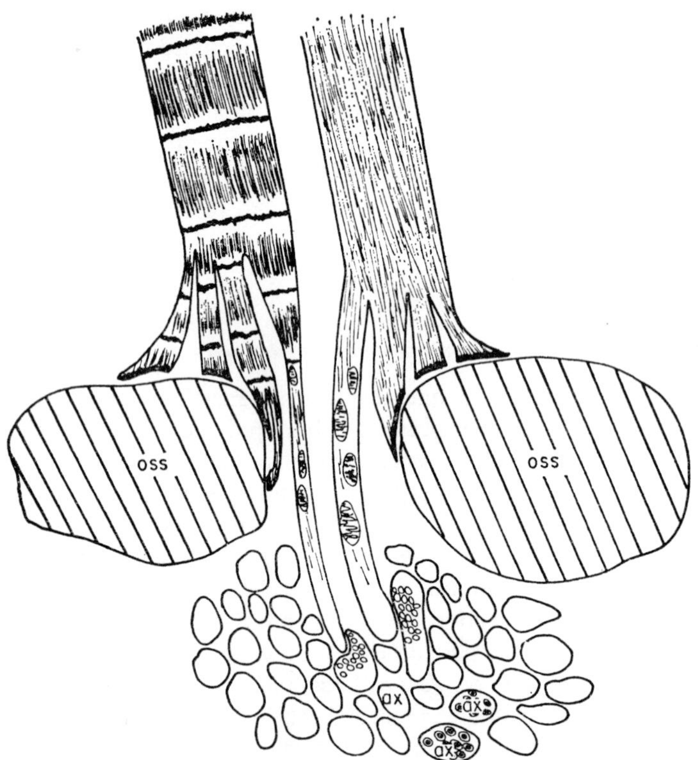

FIG. 25. Diagram of the innervation of the muscles of a tridentate pedicellaria of *Echinus*. The muscles are inserted upon the ossicles but processes from the muscles penetrate between the ossicles (*oss*) and pass into a nervous region (*ax*). Axons containing synaptic vesicles there synapse upon the processes of the muscles. Dense-cored vesicles are also present within some axons. Not to scale.

in echinoderm nerve fibres, but microtubules 260Å in diameter have recently been described in the axons of the haemal vessel of *Cucumaria frondosa*.[205] The membranes themselves appear to lack any particular specializations, as do the axo-axonic membranes. Specialized axons, called ribbon axons, were described by Smith in *Astropecten irregularis*.[722] His description was followed by further examples from *C. sykion*[633] and *Stichopus mollis*.[271] Electron microscopy has not confirmed the existence of any nerves supplying the ampullary muscles of *Asterias rubens*[39] and it seems probable that conduction is mechanical from cell to cell. Cobb has reinvestigated the ampulla of *Astropecten irregularis* with the aid of the electron microscope and has shown that nerves are indeed absent from this structure.[142] Extensions of the main soma of the muscle cells pass down the seam of the ampulla and synapse with axons from the radial nerve cord in a special area of the connective tissue bulbs found in the neck region of the tube foot ampulla system. These later terminate in the thick collagenous walls of the tube feet. These muscle extensions stain with methylene blue and take on the appearance of flat wavy structures reminiscent of the ribbon axons described by Smith, and doubtless they are one and the same thing. The synapses possess similar characteristics to those described in the hyponeural tissue of *E. esculentus*.[146]

Electrolyte Balance

Whilst little or nothing is known about the distribution of electrolytes within echinoderm nerve fibres, the position with respect to muscle fibres is a little better understood. Determination of the accurate partitioning of the ions presupposes a knowledge of the extracellular space. This was first estimated by Steinbach using histological and chloride analysis techniques upon the pharyngeal retractor muscles of *T. briareus* and found to be about 43%.[733] Later work upon the longitudinal muscle of *S. mollis* yielded a somewhat lower figure.[713] There is evidence that molecules, such as inulin, which are often used in determinations of this sort, do in fact slowly enter the muscle fibres and consequently too high a value for the extracellular space results. Recently albumin iodinated with I^{131} has been employed, thus making it possible to determine its concentration levels without having to destroy the muscle. Such experiments yield a value for the extracellular space of this muscle of 29·7%. Simultaneous measurements of Na^{22} and the iodinated albumin (RHISA) indicated that some 36% of the sodium and chloride is intracellular whereas Steinback thought that chloride was extracellular.[734] Efflux experiments with Na^{22} failed to differentiate a fast and slow component and it was concluded that the cell membrane does not impose a severe barrier to the diffusion of these ions. By comparison to a model with the appropriate dimensions, the exchange of sodium was found experimentally to be slower than could be accounted for by simple diffusion and an increased path length was postulated. However, similar experiments using K^{42} showed a clear distinction into fast and slow components.[712] The process of exchange was not in fact complete after 13 hr, and so it is probably incorrect to speak of a non-exchanging fraction. In this respect holothurian longitudinal muscle resembles vertebrate skeletal muscle. Raising the level of potassium in the bathing medium increased the exchange rate, but did not significantly affect the internal concentration of potassium, until the external level reached 120 mM/l. In vertebrates, on the other hand, the rate of entry of potassium into muscles is highest in a high potassium/low sodium ratio solution. The situation recalls the condition found in toad sartorius muscle, where the failure to fulfil the equation $K_o \times Cl_o = K_i \times Cl_i$ suggested that a Donnan equilibrium did not obtain. The data did, however, approximate to the model proposed for amphibian muscle which postulated that the muscle cylinder had a resistive coating with adsorptive properties.[322] Furthermore, in solutions where the potassium concentration had been increased but the sodium concentration had been correspondingly decreased, there was a consistent decrease in the volume of the muscle in spite of the fact that the solutions were isotonic. Such shrinkage would appear to be correlated with the resulting contracture, vertebrate muscle under these conditions undergoing swelling. Over a wide range of potassium levels—10 to 240 mM/l KCl—the ratio Cl^-_o/Cl^-_i remains nearly constant and the chloride space is not markedly altered. In vertebrate muscle under these conditions the ratio falls and the chloride space rises. Respiratory changes, especially those associated with the spontaneously active holothurian cloaca, which occur when the ionic balance of the external medium is altered, have already been referred to in Chapter 9.

Mechanical Properties and Ultrastructure

Thus the picture of the functioning of these echinoderm muscles so far investigated, both from an electrical and ionic standpoint, is not greatly at variance with that of vertebrate skeletal muscle and furthermore it differs only in its detailed behaviour. Its mechanical properties are not greatly dissimilar from those of frog skeletal muscle. A. V. Hill in his classical researches upon muscle demonstrated this point in 1926 by means of quick release experiments upon the longitudinal muscle of *Holothuria nigra*.[349] A sudden shortening of the muscle when already in isometric tension results in a fall in that tension followed by a rise to a new value appropriate to the new length. This subsequent rise in tension follows the same time course as does the rise in tension for an initial isometric contraction. Hill was led to conclude that the viscous elastic properties of the muscle were similar to those of frog skeletal muscle but that the time course was much slower. A similar conclusion was reached after further studies upon both this muscle and the lantern retractor muscles of *E. esculentus* in comparison with dogfish jaw muscle.[462] Contraction was postulated to be due to a reorganization of the molecular pattern within the muscle itself. Just what form this pattern takes has been revealed in recent years by electron microscopy. The extremely regular pattern of myofilaments seen in vertebrate striated muscle occurs in echinoderms, whose muscles are nevertheless considered unstriated. In the lantern retractor muscle of *E. esculentus* and in the jaw-closing muscles of the pedicellariae, quite a high degree of orientation of the myofilaments is to be observed, although their axes are not always uniformly parallel with the major axis of the cell.[146,147] This situation is somewhat similar to the condition found in smooth muscle with which echinoderm muscles have some physiological similarities.

Muscle Proteins

The nature of the muscle proteins of echinoderms have received scant attention. Maruyama and Matsumiya[498] prepared a contractile protein extract from the tube feet of *Asterias amurensis.* Broadly it had many of the characteristics of actomyosin preparations from higher animals but certain differences were apparent. The solubility in potassium chloride was less than for other actomyosins, and a low ATPase activity was suggestive of the inclusion of inert material such as tropomyosin A, as has been found in other invertebrates. The addition of 10 mM magnesium chloride increased the ATPase activity, whereas in all other animals examined, except the sea anemone, an inhibitory effect has been observed.

Chemical Transmitters and the Action of Drugs

The existence of chemical transmitters at the neuromuscular junctions, unlike the axo-axonic synapses so far examined, is based upon a considerable body of evidence. Such evidence is twofold. Firstly there is the demonstration that either the transmitter or its lytic enzyme will produce or modify response from nerve/muscle preparations, and secondly, the extraction of the transmitter or its associated enzyme system from the muscle itself or from the bathing medium. Otto Reisser[659] was probably the first to show that an echinoderm muscle, namely the longitudinal muscle of *Holothuria* sp., was sensitive to administration of acetylcholine at very low concentrations (5×10^{-8} g/ml). He later demonstrated that the responses to electrical stimulation of the same muscle from *H. stellata* were potentiated by eserin, a well-known anticholinesterase.[660] This work was later extended by Bacq in a series of papers.[28–34] He demonstrated the presence of an acetylcholine-like substance in extracts of longitudinal body wall musculature of *H. tubulosa*, the activity of such extracts being destroyed by alkaline hydrolysis and reaction with *Octopus* blood which was known to contain cholinesterase.[28] Additionally eserin was found to potentiate the activity of the extracts and also to sensitize a similar muscle in the related *H. stellata*, not only to acetylcholine, but also to histamine, tyramine, choline and potassium ions, a property which unfortunately rendered this muscle unsuitable as an assay preparation.[29] A similar non-specific sensitivity to eserin was also found in the lantern retractor muscles of *E. esculentus.*[31] Other workers have also found preparations from several species of holothurian to suffer similar disadvantages.[808] Values for the acetylcholine content of echinoderm tissues are recorded in Table 11.

TABLE 11. ACETYLCHOLINE CONTENT OF ECHINODERM TISSUES

Species	Tissue	μg/g ACh	Reference
HOLOTHUROIDEA			
Holothuria tubulosa	long muscle	1·5 to 1·7	Bacq [29]
	intestine	0·4	Bacq [29]
	intestine	0·8	Corteggiani quoted Welsh [809]
Parastichopus californicus	radial nerve	50·0	Roberts quoted Welsh [809]
ASTEROIDEA			
Asterias glacialis	intestine	12·0	Corteggiani quoted Welsh [809]
A. forbesi	radial nerve	72	Cottrell quoted Welsh [809]
Pteraster tesselatus	radial nerve	100	Roberts quoted Welsh [809]
Pisaster ochraceous	radial nerve	100	
Pycnopodia helianthoides	radial nerve	100	
Leptasterias hexactis	radial nerve	100	
ECHINOIDEA			
Paracentrotus lividus	testis	0	Corteggiani quoted Welsh [809]
	ovary	0·08	
Strongylocentrotus droebachiensis	radial nerve	40	Roberts quoted Welsh [809]

A convincing piece of evidence regarding the cholinergic nature of neuromuscular transmission was the demonstration that upon repeated electrical stimulation, the eserinized longitudinal muscle of *H. tubulosa* released into the bathing medium a substance which behaved exactly like acetylcholine and which could be assayed upon the classical leech dorsal muscle preparation. The longitudinal muscle of *S. regalis* has also been employed as an assay preparation for acetylcholine,[9,32,694,696] and also that of *H. grisea* which in general does not display the irregular and spontaneous activity of other species. A quite incredible sensitivity to acetylcholine down to concentrations of the order of 1×10^{-14} g/ml has been recorded. However, not all workers have confirmed the effects of acetylcholine and it was found to be without effect upon the cloacal complex of *Holothuria* sp.[222] The potentiation of many of these responses by eserin points very strongly to a cholinergic system in such muscles as the longitudinal muscle in *S. californicus*[366] and, rather surprisingly in view of the above-mentioned results, the isolated holothurian cloaca.[84] A slow gain in tension and often a development of rhythmicity in the lantern retractor muscles of *Parechinus angulosus* has been recorded after treatment with eserin.[68] Bacq first demonstrated the presence of a cholinesterase system in echinoderms[30] and these observations were extended to include whole body extracts of *Antedon bifida* and *Asterina gibbosa* and of the lantern retractor muscles of *E. esculentus*.[31] Table 12 gives the results of surveys by other workers (see also review by Florey[248]).

TABLE 12

Species	Tissue	mg ACh hydrolyzed per 100 mg tissue	Reference
HOLOTHUROIDEA			
Holothuria nigra	long muscles	0·037	34
Thyone briareus	long muscles	0·820	100
Cucumaria lactea	whole body	1·517	24
Mesothuria intestinalis	whole body	0·308	24
ASTEROIDEA			
Asterias rubens	intestine and stomach	0·462	24
	ampullae and podia	0·194	24
A. forbesi	radial nerve cord	0·384	100
OPHIUROIDEA			
Amphiura chiajei	whole body	0·152	24
ECHINOIDEA			
Psammechinus miliaris	intestine	0·508	24
Echinus esculentus	intestine	1·369	24
Paracentrotus lividus	plutei	0·700	25

The case for the operation of a cholinergic type of transmission at the neuromuscular junction of many preparations therefore rests upon the following observations. Firstly the demonstration of both acetylcholine and cholinesterase within the muscle and the potentiation of responses by eserin, a known anticholinesterase. Secondly, the release of an acetylcholine-like substance into the bathing medium of a preparation which has been repeatedly stimulated electrically. As already noted, the case for cholinergic transmission at the axo-axonic synapses is less well established. Conduction in the radial nerve cord of *Asterias rubens* is abolished by treatment with M/1000 eserin.[62a] Acetylcholine has been detected in the radial nerve cords of a number of echinoderms and the cholinesterase system recorded from one of them. The presence of what are undoubtedly vesicles at these synapses is important circumstantial evidence and would warrant further investigation.

In many groups of animals, chiefly the vertebrates, cholinergic transmission can be referred to two distinct categories depending upon whether the effects of acetylcholine can be mimicked by nicotine or muscarine. The former are most usually referable to ganglia and skeletal muscle, whilst the latter is more usually associated with smooth and cardiac transmission. The results of a number of workers who have tried the effects of a wide range of drugs upon echinoderm preparations of nerves and muscles are to be found in Table 13, listed with these categories in mind. Regrettably this work has only been caried out with a comparatively few species of holothurian and a single species of echinoid. The evidence is somewhat conflicting, but two main types of preparation have been employed. These may be loosely termed skeletal, such as the longitudinal body wall muscles, and visceral, such as the holothurian cloaca. Reference to Table 13 shows that the skeletal preparations have a predominantly nicotinic reaction, but that the visceral preparations are less conclusively muscarinic. However, too great an emphasis upon this distinction should not be made, for in many cases the histology of the preparations used was poorly known. Because of their small size, a fine nerve plexus may go unrecognized and thus to relate the neuromuscular organization of echinoderms to that of higher animals upon the basis of drug action is somewhat premature and more work at the histological level is required. The problem of the specificity of drug action, particularly acetylcholine, has been studied in relation to the pharyngeal retractor muscles of *C. frondosa*. Choline specificity is low and contractions may be induced by many nicotinomimetics and muscarimimetics.[481,482] Specificity

TABLE 13. NICOTINIMIMETIC AND MUSCARIMIMETIC ACTIONS OF ACETYLCHOLINE UPON ECHINODERM NERVE MUSCLE PREPARATIONS

Nicotinic	Ref.	Muscarinic	Ref.
1. NICOTINIC		1. MUSCARINIC	
a. *Stimulated by nicotine*		a. *Stimulated by muscarine*	
Rise in tone of rhythmic cloaca of *Stichopus moebii.*	814	No information.	
		b. *Stimulated by pilocarpine*	
Contracture of longitudinal muscle of *S. regalis.*	32	Without effect upon cloaca of *Stichopus moebii.*	814
Decreased rate and amplitude of longitudinal muscle of *Stichopus californicus*, also contracture.	366	Without effect upon longitudinal muscle of *S. californicus.*	366
Contracture of longitudinal muscle of *Holothuria grisea.*	696 695		
Activation of quiescent lantern of *Parechinus angulosus.*	69		
2. NICOTINIC BLOCKING AGENTS		2. MUSCARINIC BLOCKING AGENTS	
a. *Nicotine*—in certain circumstances.		a. *Atropine*	
Inhibits haemal vessel of *Parastichopus californianus.*	644	Blocks action of ACh on longitudinal muscle of *Stichopus regalis.*	32
b. *Curare*		Inhibits cloaca of *S. moebii* at high conc.	814
ACh induced contractions of longitudinal muscle of *Thyone briareus* NOT blocked.	209	Slows amplitude, rate and tonus of longitudinal muscle of *S. californicus.*	366
ACh induced contractions of longitudinal muscle of *Holothuria grisea* ARE blocked.	9	Slows beat of dorsal sinus of *S. moebii.*	814
Spontaneous contractions of pharyngeal retractor muscles of *Cucumaria sykion* abolished	635		
Increases tone of lantern retractor muscles of *Parechinus angulosus*, will activate quiescent preparations, but high concentrations inhibitory.	69		

increases with phylogenetic advance and it is suggested that this is accompanied by restriction of the size of the cholinoceptic zone upon the muscle membrane.

Numerous other drugs have been tried upon echinoderm nerve muscle preparations or whole animals. Once again the results are variable and often conflict, not surprisingly, with the established activity for the drug upon higher animals. Such experiments range from the reversal of certain types of starfish behaviour by strychnine[545] to the extensive pharmacological exercises by Iriye and Dille.[366] One result of such work has been to reveal the possible participation of chemical transmitters other than acetylcholine. The pharmacology of the holothurian cloaca has already been described in Chapter 9 and it will be recalled that adrenalin was found to have an inhibitory effect upon this structure.[814] Prior treatment with other drugs such as cocaine abolished the inhibitory effect and resulted in an excitatory action. This situation is somewhat reminiscent of the mammalian gut, but once again direct comparison at the neuromuscular level would be inappropriate. Noradrenalin has an inhibitory effect upon the lantern retractor muscles of *P. angulosus*[68] but an excitatory response was recorded from the longitudinal muscle of *S. californicus*.[366] The beat of the dorsal sinus of *S. moebii* was also accelerated by adrenalin.[814] Interesting antagonistic behaviour between adrenalin and acetylcholine, again suggestive of a dual innervation reminiscent of the higher animals, resulted from the finding that treatment with 10^{-4}M adrenalin relaxed an acetylcholine-induced contraction of the longitudinal muscle of *T. briareus*.[209] A very elegant demonstration of such antagonism was made by Anderson,[11] who showed that the mouth of the starfish *A. forbesi* could be opened, due to contraction of the radial muscles, upon application of acetylcholine to the peristomial membrane. Closure of the mouth, due to contraction of the circular muscles, followed application of adrenalin. Antagonistic effects have also been observed upon the fibres of the stomach retractor muscles. These muscles are divided into intrinsic and extrinsic portions, the first of these being unaffected by drugs. The extrinsic strands contract when irrigated with acetylcholine and relax when washed, but will relax more rapidly if washed with adrenalin.[102] These extrinsic fibres also contract when stimulated electrically and exhibit slight summation with successive shocks at 2-sec intervals.[43] Regeneration of the intrinsic muscle strands and the flagellated gutter patterns requires re-establishment of contact with the extrinsic retractor nodule. Failure to do so results in marked inhibition of regeneration.[16] The lantern retractor muscle of *P. angulosus* is also relaxed from an acetylcholine contracture by application of adrenalin. Other examples of this type, such as the holothurian cloaca and the haemal vessels, are dealt with in Chapter 9. The longitudinal muscle of *H. stellata* and *H. tubulosa* was rendered hyperexcitable to the action of adrenalin by prior treatment with eserin,[32] but adrenalin was found to have an inhibitory effect upon the cloacal complex of *S. moebii*.[222,814]

Attempts to demonstrate the presence of adrenalin or an adrenalin-like substance in echinoderms has not, at least until recently, met with such conclusive results as has those with acetylcholine. Neither adrenalin nor noradrenalin were to be found in *Mesothuria intestinalis*, *E. esculentus* or *Ophiura albida*, but the whole body extracts did produce a slight fall in the blood pressure of a cat and had a relaxing effect upon the fowls rectum.[608] The inactivator of adrenalin corresponding to cholinesterase is amine oxidase and the presence of this enzyme has been established in the digestive glands of four species of asteroid and in the gut of two echinoids, but not in any of the species of holothurian so far examined. Recently Cottrell, using more refined techniques of fluorescence chromatography, has estimated the noradrenalin content of the radial nerve cords of *E. esculentus* and *A. rubens*, at between 0·5 and 3·5 μg/g wet weight and a dopamine content of 2·5 to 8·0 μg/g wet weight.[159] When intact animals were allowed to remain for 5 days in sea water containing 3 μg/ml reserpine, a potent inhibitor of amine oxidase, the noradrenalin content was depleted by between 20 and 50% and the dopamine content by 50 to 70%. It was interesting to find that after such treatment the animals took considerably longer to right themselves which is reminiscent of the depressant action of noradrenalin upon the lantern retractor muscle of *Parechinus* referred to above. However, the speed of conduction seems to be little affected by treatment of the radial nerve cord with reserpine.[62b]

Other chemical transmitters have also been sought in echinoderms. Mammalian brain extracts contain a principle with properties antagonistic to acetylcholine, Factor 1, and is thought by some authorities to be identical to γ-amino butyric acid (GABA). Contraction of the longitudinal muscle of *Actinopyga agassizi* recorded after application of these extracts may possibly have been due to the presence of acetylcholine itself as an impurity.[247] Paralysis of the spinal musculature of *D. antillarum* resulting from spraying the extract on to the inside of the test may be due to its action upon the radial nerve cord rather than the musculature directly. There is evidence that Factor 1 and acetylcholine compete for the same receptor sites and the exact nature of those sites may be as important as the nature of the transmitter. Acetylcholine-induced contractions of the oesophagus of *Strongylocentrotus droebachiensis* were inhibited with extracts containing Factor 1, but not with GABA. Such evidence suggests that Factor 1 and GABA are not in fact synonymous[249] and GABA itself does not appear to have been isolated from echinoderms. Another substance with effects antagonistic to acetylcholine is 5-hydroxytryptamine. In the mollusc *Mytilus edulis* it brings about relaxation of an acetylcholine-induced contraction of the anterior byssus retractor muscle. In an examination covering one species of asteroid, echinoid and holothurian, only minute quantities of 5-HT were found in whole body homogenates.[810] The circumoral region of *T. briareus*, which of course contains the bulk of the animal's nervous tissue, contains only a similarly low level. However, the presence of quantities of the order of tenths of a μg/g wet weight have been demonstrated in the radial nerve cord of the starfish *Asterias forbesi*.[809] The occurrence of this neurohumoural substance in non-nervous echinoderm tissues is also noted.

Photic Stimulation of Nerve/Muscle Systems

Apart from electrical and chemical stimulation so far considered, one further form of excitation must now be added. In 1963 Boltt and Ewer succeeded in eliciting various and repeatable responses from the lantern retractor muscle of *P. angulosus* to illumination.[68] Preparations which did not display spontaneous activity in the dark would contract after a 20 sec exposure to light. The ensuing contraction would in many cases exhibit a quick and a delayed response. Very brief flashes of light, of the order of 10^{-3} sec, would still elicit the quick response. Generally the magnitude of this contraction was increased with increasing intensity of illumination, if the duration was kept constant, the converse being also true. The latent period also decreased, prolonged illumination often resulting in the development of spontaneous activity. Repetitive stimulation showed that the initial response was greater than for subsequent stimuli, such an effect being maximal with intervals of 300 sec. If the second light stimulus be a high intensity one, the contracted muscle could be made to relax somewhat. Experiments to find the wavelength with maximal inhibitory effect failed, but it was suggested that both excitatory maxima and inhibitory maxima were in the blue-green region of the spectrum. Exactly which structures were being stimulated by the light in this research is none too clear. Such complex behaviour is suggestive that some neural centre is involved and Ewer now inclines to the view that the muscle cells may be acting as light guides transmitting the stimulus to some more remote part of the preparation.[144] Stimulation of the antagonistic comminator muscles results in a loss of tone by the lantern retractors in a manner similar to that found in *C. sykion*.[635] In this animal the pharyngeal retractors are antagonized by the longitudinal muscles of the body wall and stretching them resulted in the abolition of the spontaneous activity of the pharyngeal retractors, whereas in *Parechinus* it was the quick and delayed contractions which were extinguished. Inhibitory effects of light are well known for a number of invertebrate preparations. Maximum inhibition in the shadow response in *D. antillarum* was about 455 $m\mu$,[836] but the pigment involved has not been identified, for the *in vivo* absorption maximum of naphthoquinone is in the region of 550 $m\mu$.[531] Examples of the actual inhibition of nervous activity by light have not been observed in echinoderms, except for the "off" effect in *Diadema* radial nerve already discussed.[748] Better examples are to be found in other phyla and include for example inhibition in the pallial nerve of the lamellibranch *Spisula* and the decapod *Procambarus*.[395,396] The wavelength at which inhibition is maximal is around 540 $m\mu$ and is thought to indicate the participation of a carotenoprotein. The most effective wavelengths producing excitation in a blue adapted preparation are in excess of 600 $m\mu$. The possibility that a haemoprotein present in the neurons in a high concentration is reduced by the lack of sensitivity of this response at wavelengths around 400-420 mμ, the Soret band for a haemoprotein. Presumably this would only be true if it were the protein component which was responsible for absorbing the light.

Learning

The general reactions of echinoderms to a variety of stimuli including light have been considered in Chapter 7, and it was concluded that these were rather stereotyped, and, normally, the more intense the stimulus, the more intense the response. The characteristics of the neuromuscular system outlined in this chapter indicate, with certain reservations, a rather unsuspected complexity, both in terms of organization and physiology. It is obvious that the comparisons often made with the coelenterates are applicable only to superficial responses and not to their behaviour as a whole.[99] The suspected pacemaker activity of the radial nerve cords would be such an example. However, synaptic regions appear to lack membrane specialization and true facilitation has yet to be demonstrated at either the axo-axonic junctions or neuromuscular junctions. Even allowing for such limitations, it is rather surprising that echinoderms possess little or no ability to modify their behaviour in the light of past experience. There is little modern published work upon this aspect, either behavioural or upon the effects of ablation or extirpation of the nervous system upon their learning capacities.

As far back as 1886, Preyer claimed that he could detect some ability to learn in asteroids and ophiuroids. His experiments were confined to allowing the animals to free themselves from fetters or to removing a rubber tube placed over one of the arms. A seemingly increasing facility at this performance was thought to indicate a degree of learning and the possession of some intelligence.[640] Experiments with *A. rubens* produced similar conclusions.[787] Preyer's experiments were criticized on the grounds that there was no increase in the frequency of the various actions performed by the animals in order to extricate themselves, nor was the total time required reduced.[304] The evaluation of such parameters was made more difficult by the fact that ophiurans in particular pass from one level of activity to another with great speed, thus permitting an active animal to give the impression of being an intelligent one. Many echinoderms have a well-developed tendency to right themselves on particular radii and when moving about appear to have a distinct bilateral symmetry with a physiologically dominant anterior end.[406,407] This characteristic was employed in *Astrometis sertulifera* to train the animals to turn on different arms.[371] It was found that they could retain this memory for up to 5 days. Ophiuroids were found to possess the elements of what would now be called a kinaesthetic sense, in that the animals remembered contact stimuli with the walls of their aquarium.[163] It was, however, very transitory and was quickly submerged by stronger stimuli. Diebschlag was the first to try to produce conditioned reflexes in echinoderms.[199] He trained *Astropecten bispinosus*, *A. spinulosus* and *Ophiothrix fragilis* to draw back from the boundary between rough and smooth, ribbed and plain and dark and light surfaces. As punishment he administered asteroid pedicillariae to the animal's arms. His results too suggested the existence of a brief memory. More recently attempts have been made to develop conditioned reflexes to tactile stimuli, namely, the nature of the substrate upon which the animal was moving.[727] Instead of punishment the reflex was reinforced by the presentation of food. After 4 to 9 such presentations a conditioned reflex developed which was stabilized after a further 8 to 14 presentations. Regrettably its persistence does not appear to have been investigated. *Pisaster giganteus* when kept in an aquarium tends to cling to the walls, descending to the floor to retrieve food placed there and return up the wall to feed upon it.[443] Individuals were trained to descend to the floor of the tank when it was illuminated, a mussel being presented as a reward. After 8 such trials, tests were made in which the mussel was not presented until 15 min after the illumination so as to preserve the association. This was continued for some 9 to 18 tests, during which period all the animals responded by commencing their descent before the 15 min period had elapsed. After these tests, no mussels were presented at each test, and gradually fewer and fewer animals responded, and eventually ceased altogether.

Such limited learning ability may be sufficient to keep the animals within the confines of a pier from which a rain of *Mytilus* descended upon them.[443] Although many thousands of starfish were to be found under such a pier, they did not extend more than 5 ft on either side of it. Destruction of the circumoral nerve ring in a starfish with a stabilized conditioned reflex led to an abrupt disturbance of the reflex, but it was restored after 2 to 3 months without further training following the regeneration of the nervous tissue.[727] A second destruction resulted in permanent loss of the reflex and it was concluded that although the circumoral nerve ring played a decisive role in the formation of a conditioned reflex, the radial system too has some capability in this way. In view of earlier work where the capacity to retain a reflex for more than a few days was absent, it seems remarkable that these animals should have retained a conditioned reflex for some months without further training and be capable of passing it on to regenerating tissue. Although echinoderms exhibit a wide range of activities in response to their environment at large and to specific stimuli in particular, their potentialities are somewhat limited due to restriction of their neural capacity in terms of memory and learning. To some extent this is a reflection of their radial symmetry with its associated absence of a discrete head and attendant brain or ganglia. The complex nature of the neuropil within the small aggregates of nervous tissue which may be referred to as ganglia in other parts of the body has already been demonstrated, but is obviously insufficient for any higher nervous processes.

REFERENCES

1. Abeloos, M., 1926. Sur les pigments tégumentaires des Astéries. *C. r. Séanc. Soc. Biol.* **94,** 19-21.
2. Abeloos, M. and Teissier, G., 1926. Notes sur les pigments animaux. *Bull. Soc. zool. Fr.* **51,** 145-151.
3. Abraham, M., 1963. Etudes morphologiques sur les coelomocytes des Échinides reguliers. *Israel J. zool.* **12,** 101-116.
4. Abraham, M., 1964. La coagulation dans le liquide periviscéral des échinides etudiée à l'aide du microscope à contrast de phase. *Pubbl. Staz. zool. Napoli* **34,** 43-52.
5. Ackermann, D., 1935. Asterubin: eine schwefelhaltige Guanidinverbindung de belebten Natur. *Hoppe-Seyler's Z. physiol. Chem.* **232,** 206-212.
6. Ackermann, D. and Hoppe-Seyler, G., 1963. Asterit: eine neuer biologischer Cyclit. *Naturwissenschaften* **50,** 733-734.
7. Allen, W. V., 1965. Lipogenesis in the sea star *Pisaster ochraceous. Diss. Abstr.* **26,** 454.
8. Allen, W. V. and Giese, A. C., 1966. An *in vitro* study of lipogenesis in the sea star *Pisaster ochraceous. Comp. Biochem. Physiol.* **17,** 23-38.
9. Ambache, N. and Sawaya, P., 1953. Use of *Holothuria grisea* for acetylcholine assays of electric organ extracts from *Narcine braziliensis* (Älfers). *Physiologia comp. Oecol.* **3,** 53-56.
10. Anderson, J. M., 1953. Structure and function in the pyloric caeca of *Asterias forbesi. Biol. Bull. mar. biol. Lab. Woods Hole* **105,** 47-61.
11. Anderson, J. M., 1954. Studies on the cardiac stomach of the starfish *Asterias forbesi. Biol. Bull. mar. biol. Lab. Woods Hole* **107,** 157-173.
12. Anderson, J. M., 1959. Studies on the cardiac stomach of a starfish *Patiria miniata* (Brandt). *Biol. Bull. mar. biol. Lab. Woods Hole* **117,** 195-201.
13. Anderson, J. M., 1960. Histological studies on the digestive system of a starfish *Henricia,* with notes on Tiedemann's pouches in starfishes. *Biol. Bull. mar. biol. Lab. Woods Hole* **119,** 371-398.
14. Anderson, J. M., 1962a. Pyloric caeca and Tiedemann's pouches in *Linckia columbriae. Am. Zool.* **2,** 387.
15. Anderson, J. M., 1962b. Studies on visceral regeneration in sea stars. I. Regeneration of pyloric caeca in *Henricia leviuscula* (Stimpson). *Biol. Bull. mar. biol. Lab. Woods Hole* **122,** 321-342.
16. Anderson, J. M., 1965. Studies on visceral regeneration of sea stars. III. Regeneration of the cardiac stomach of *Asterias forbesi. Biol. Bull. mar. biol. Lab. Woods Hole* **129,** 454-470.
17. Anderson, R. S., 1966. Anal pores in *Leptosynapta clarki* (Apoda). *Can. J. Zool.* **44,** 1031-1035.
18. Andrew, W., 1962. Cells of the blood and coelomic fluid of tunicates and echinoderms. *Am. Zool.* **2,** 285-297.

19. ANTON, H., 1927. Die Ko-ordination der Saugfüsschenreflexe der Regulären Echiniden. *Z. vergl. Physiol.* **5,** 801-816.
20. ARAKI, G. S., 1965. On the physiology of feeding and digestion in the sea star *Patiria miniata. Diss. Abstr.* **25,** 4306.
21. ARVANITAKI, A. and CHALAZONITIS, N., 1949. Inhibition aux excitation des potentiels neuroniques à la photoactivation distincte de deux chromoprotéides (Caroténoides et chlorophyllien). *Arch. Sc. Physiol.* **3,** 45.
22. ARVY, L., 1954. Toxicité des tissues de *Holothuria tubulosa,* de *H. poli, et de H. impatiens* pour quelques vertebrés. *C. r. hebd. Seanc. Acad. Sci. Paris* **239,** 1432-1434.
23. AUGUSTINSON, K.-B., 1946a. Cholinesterase and its specificity. *Nature, Lond.* **157,** 587.
24. AUGUSTINSON, K.-B., 1946b. Cholinesterase in some marine invertebrates. *Acta physiol. scand.* **11,** 141-150.
25. AUGUSTINSON, K.-B., 1948. Cholinesterases. *Acta physiol. scand.* **15,** Suppl. 52, 1-182.
26. AWERINZEW, S., 1911. Ueber die Pigmente von *Strongylocentrotus droebachiensis. Archs. Zool. exp. gen.* Ser. 5 **8,** i-viii.
27. BACESCO, M. C. and MARGINEAU, C., 1959. Éléments Méditerranéens Bouveaux dans la faune de la mer Noire, rencontrés dans les eaux de Roumélie (Norde Ouest Bosphore). Données nouvelles sur le problème du peuplement actuel de la mer Noire. *Arch. Oceanogr. Limnol.* **11,** 63-74.
28. BACQ, Z. M., 1935a. Occurrence of unstable choline esters in invertebrates. *Nature, Lond.* **136,** 30-31.
29. BACQ, Z. M., 1935b. Recherches sur la physiologie et la pharmacologie, du système nerveux autonomie. XVII. Les esters de la choline dans les extraites de tissus des invertébrés. *Archs. int. Physiol.* **42,** 24-42.
30. BACQ, Z. M., 1935c. Recherches sur la physiologie et la pharmacologie du système nerveux autonomie. XIX. La choline esterase chez les invertébrés l'insensibilitié des Crustacés a l'acetylcholine. *Archs. int. Physiol.* **42,** 47-60.
31. BACQ, Z. M., 1937. Nouvelles observations sur l'acetylcholine et la choline esterase chez les invertebres. *Archs. int. Physiol.* **44,** 174-189.
32. BACQ, Z. M., 1939a. Un test marin pour l'acetylcholine. *Archs. int. Physiol.* **49,** 20-24.
33. BACQ, Z. M., 1939b. Action de l'eserine chez les Holothuries et chez les Ascidies presence de nerfs cholinergiques chez les Holothuries. *Archs. int. Physiol.* **49,** 25-32.
34. BACQ, Z. M. and NACHMANSOHN, D., 1937. Cholinesterase in invertebrate muscles. *J. Physiol. Lond.* **89,** 368-371.
35. BALDWIN, E. and NEEDHAM, D. M., 1937. A contribution to the comparative biochemistry of muscular and electrical tissues. *Proc. R. Soc.* B. **122,** 197-219.
36. BALDWIN, E. and YUDKIN, W. H., 1950. The annelid phosphagen with a note on phosphagen in Echinodermata and Protochordata. *Proc. R. Soc.* B. **136,** 614-631.
37. BAMBER, R. C., 1921. Notes on some experiments on the water vascular system of *Echinus. Proc. Trans. Liverpool Biol. Soc,* **35,** 64-70.
38. BANG, F. B. and LEMMA, A., 1962. Bacterial infection and reaction to injury in some echinoderms. *J. Insect Path.* **4,** 401-414.

39. Bargmann, W. and Behrens, Br., 1963. Uber den feinbau des Nervensystems des Seesternes (*Asterias rubens* L.). II. Zur frage des baues und der innervation der ampullen. *Z. Zellforsch. mikrosk. Anat.* **59,** 746-770.
40. Barnes, E. W., 1946. Starfish menace in Southern Massachusetts, 1931. *Bull. Bingham oceanogr. Coll.* **9,** Art. 3, 38-43.
41. Barnes, J. and Endean, R., 1964. A dangerous starfish *Acanthaster planci. Med. J. Aust.* **51,** 592-593.
42. Barthels, P., 1895. Notiz uber die Excretion der Holothurien. *Zool. Anz.* **18,** 493-494.
43. Basch, P. F., 1956. Observations on the retractor strands of the starfish stomach. *Biol. Rev. City Coll. New York* **18,** 14-17.
44. Bather, F. A., 1900. *Echinodermata. A Treatise on Zoology,* Part III. Ed. E. R. Lankester. London, Black, 344 pp.
45. Benazzi-Lentati, G., 1941. Sulla distribuzione del glicogeno e sulla glicemia vera degli invertebrati. *Archo. zool. ital.* **29** (suppl.), 35-69.
46. Bennett, J. and Giese, A. C., 1955. The annual reproductive and nutritional cycles in two western sea urchins. *Biol. Bull. mar. biol. Lab. Woods Hole* **109,** 226-237.
47. Berger, E. and Bethe, A., 1931. Die Durchlassigheit der Körperoberfäschen wirbelloser Tiere fur Jodionen. *Pflugers Arch. Physiol.* **228,** 768-789.
48. Bergmann, W., 1949. Comparative biochemical studies on the lipids of marine invertebrates, with special reference to the sterols. *J. mar. Res.* **8,** 137-176.
49. Bergmann, W., 1963. Sterols: their structure and distribution, in *Comparative Biochemistry,* Ed. Florkin and Mason, Academic Press, London.
50. Bergmann, W. and Stansbury, H. A., 1943. Astrol. *J. org. Chem.* **8,** 283-284.
51. Bergmann, W. and Stansbury, H. A., 1944. The sterols of starfish. II. *J. org. Chem.* **9,** 281-289.
52. Berrill, M., 1966. The ethology of the synaptid holothurian *Opheodesoma spectabilis. Can. J. Zool.* **44,** 457-482.
53. Bertolini, F., 1933. Sulle funzioni dei polmoni acquatici delle Oloturie. *Pubbl. Staz. zool. Napoli* **13,** 1-11.
54. Bertolini, F., 1934. Nuove ricerche sulla funzioni respiratoria dei polmoni acquatici delle Olothurie. *Archs. zool. ital.* **20,** 579-590.
55. Bertolini, F., 1937. L'excrezione delle Oloturie. *Proc. 12th int. Congr. Zool.* **2,** 759-760.
56. Bethe, A., 1934. Die Saltz- und Wasser-Permeabilität der Körperoberfläschen verscheidener Seetiere in ihrem gegenseitigen Verhaltnis. *Pfluger's Arch. Physiol.* **234,** 629-644.
57. Bethe, A. and Berger, E., 1931. Variatonen im Mineralbestand verscheidner Blutarten. *Pfluger's Arch. Physiol.* **227,** 571-584.
58. Bialaszewicz, K., 1933. Contribution a l'étude de la composition minerale des liquides nourriciers chez le animaux marins. *Archs. int. Physiol.* **36,** 41-53.
59. Bierry, H. and Gouzon, B., 1937. Spectre de fluorescence d'un pigment isolé des Holothuries. *C. r. Séanc. Soc. Biol.* **124,** 323.
60. Binyon, J., 1961. Salinity tolerance and permeability to water of the starfish *Asterias rubens* L. *J. mar. biol. Ass. U.K.* **41,** 161-174.
61. Binyon, J., 1962. Ionic regulation and mode of adjustment to reduced

salinity of the starfish *Asterias rubens* L. *J. mar. biol. Ass. U.K.* **42,** 49-64.

62. BINYON, J., 1964. On the mode of functioning of the water vascular system of *Asterias rubens* L. *J. mar. biol. Ass. U.K.* **44,** 577-588.

62a. BINYON, J. and HASLER, B., 1970. Electrophysiology of the starfish radial nerve cord. *Comp. Biochem. Physiol.* **32,** 747-753.

63. BLUMER, M., 1960. Pigments of a fossil echinoderm. *Nature Lond.* **188,** 1100-1101.

64. BLUMER, M., 1965. Organic pigments: their long term fate. *Science, N.Y.* **149,** 722-726.

65. BOCK, K. J. and SCHLIEPER, C., 1953. Uber den Einfluss des Saltzgehaltes im Meerwasser auf den Grundumsatz des Seesternes *Asterias rubens* L. *Kieler Meeresforsch.* **9,** 201-212.

66. BOHN, G., 1908. Sur le role et la protection des organes des sens chez les echinodermes. *C. r. Séanc. Soc. Biol.* **64,** 277-280.

67. BOLIN, L., 1926. Der geotropismus von *Psammechinus miliaris. Int. Revue ges. Hydrobiol. Hydrogr.* **16,** 125-129.

68. BOLTT, R. E. and EWER, D. W., 1963a. Studies on the myoneural physiology of the Echinodermata. IV. The lantern retractor muscle of *Parechinus:* responses to stimulation by light. *J. exp. Biol.* **40,** 713-726.

69. BOLTT, R. E. and EWER, D. W., 1963b. Studies on the myoneural physiology of the Echinodermata. V. The lantern retractor muscle of *Parechinus*: responses to drugs. *J. exp. Biol.* **40,** 727-733.

70. BOLKER, H. I., 1967a. Phylogenetic relationships of Echinoderms: biochemical evidence. *Nature, Lond.* **213,** 904-905.

71. BOLKER, H. I., 1967b. Crinosterol: a unique sterol from a comatulid crinoid. *Nature, Lond.* **213,** 905-906.

72. BONHAM, K. and HELD, E. H., 1963. Ecological observations on the sea cucumbers *Holothuria atra* and *H. leucospilotica* at Rongelap Atoll, Marshall Isls. *Pacific Sci.* **17,** 305-314.

73. BOOKHOUT, C. G. and GREENBURG, N. D., 1940. Cell types and clotting reactions in the echinoid *Mellita quinquesperforata. Biol. Bull. mar. biol. Lab. Woods Hole* **79,** 309-320.

74. BOOLOOTIAN, R. A., 1960. The effects of temperature on gonadal growth of *Strongylocentrotus purpuratus. Anat. Rec.* **137,** 342-343.

75. BOOLOOTIAN, R. A., 1961. Physical properties and chemical composition of perivisceral fluid: Echinodermata, in *Biological Handbook: Blood and Other Body Fluids.* Fedn. Am. Socs. exp. Biol. Washington, pp. 339-344.

76. BOOLOOTIAN, R. A., 1962. The perivisceral elements of echinoderm body fluids. *Am. Zool.* **2,** 275-284.

77. BOOLOOTIAN, R. A., 1963. Response of the testes of purple sea urchins to variations in temperature and light. *Nature, Lond.* **197,** 403.

78. BOOLOOTIAN, R. A., 1964. A histological study of the food canal of *Strongylocentrotus franciscanus. Helg. wiss. Meeresuntersuch.* **11,** 118-127.

79. BOOLOOTIAN, R. A. and CAMPBELL, J. L., 1964. A primitive heart in the echinoid *Strongylocentrotus purpuratus. Science, N.Y.* **145,** 173-175.

80. BOOLOOTIAN, R. A. and CAMPBELL, J. L., 1966. The axial gland complex. *Nature, Lond.* **212,** 946-947.

81. BOOLOOTIAN, R. A. and CANTOR, M. H., 1965. A preliminary report on respiration, nutrition and behaviour of *Arbacia punctulata. Life Sci.* **4,** 1567-1571.

82. BOOLOOTIAN, R. A. and GIESE, A. C., 1959a. Clotting of echinoderm coelomic fluid. *J. exp. Zool.* **140,** 207-229.

83. BOOLOOTIAN, R. A. and GIESE, A. C., 1959b. The effect of latitude on the reproductive activity of *Strongylocentrotus purpuratus. Int. oceanogr. Congr. Am. Ass. Adv. Sci.* 1959, pp. 216-217.

84. BOOLOOTIAN, R. A., GIESE, A. C., TUCKER, J. S. and FARMANFARMAIAN, A., 1959. A contribution to the biology of a deep sea echinoid *Allocentrotus fragilis. Biol. Bull. mar. biol. Lab. Woods Hole* **116,** 262-272.

85. BOOLOOTIAN, R. A. and LASKER, R., 1964. Digestion of brown algae and the distribution of nutrients in the purple sea urchin *Strongylocentrotus purpuratus. Comp. Biochem. Physiol.* **11,** 273-289.

86. BOTTAZZI, F., 1897. La pressione osmotiques du sang des animaux marins. *Archs. ital. Biol.* **28,** 61-72.

87. BOTTAZZI, F., 1906. Sulla regolazione della pressione osmotica negli organismi animali. *Arch. Fisiol.* **3,** 416-446.

88. BOTTAZZI, F., 1908. Osmotischer druck und electrische Leitfahigheit der Flussigheiten der einzelligen pflanzlichen und tierischen Organismen. *Ergebn. Physiol.* **7,** 161-402.

89. BOTTICELLI, C. R., HISAW, F. L. and WOTIZ, H. H., 1960. Estradiol 17β and progesterone in ovaries of starfish *Pisaster ochraceous. Proc. Soc. exp. Biol. Med.* **103,** 875-877.

90. BOTTICELLI, C. R., HISAW, F. L. and WOTIZ, H. H., 1961. Estrogens and progesterone in the sea urchin (*Strongylocentrotus franciscanus*) and pecten (*Pecten hericius*). *Proc. Soc. exp. Biol. Med.* **106,** 887-889.

91. BRATTSTRÖM, H., 1941. Studien über die Echinodermen des Gebietes zwischen Skaggerak und Östsee, besonders des Öresundes mit einer Übersicht über die physiche Geographie. *Unders. over Oresund.* **27,** 1-329.

92. BROOKBANK, J. W. and WHITELEY, A. H., 1954. Urease in sea urchin embryos. *Biol. Bull. mar. biol. Lab. Woods Hole* **107,** 57-63.

93. BUCHANAN, J. B., 1962. A re-examination of the glandular elements in the tube feet of some common British Ophiuroids. *Proc. zool. Soc. Lond.* **138,** 645-650.

94. BUCHANAN, J. B., 1963. Mucus secretion within the spines of ophiuroid echinoderms. *Proc. zool. Soc. Lond.* **141,** 251-259.

95. BUCHANAN, J. B. and WOODLEY, J. D., 1963. Extension and retraction of the tube feet of ophiuroids. *Nature, Lond.* **197,** 616-617.

96. BUDDINGTON, R. A., 1937. The normal spontaneity of movement of the respiratory muscles of *Thyone briareus. Physiol. Zool.* **10,** 141-155.

97. BUDDINGTON, R. A., 1942. Ciliary transport system in *Asterias forbesi. Biol. Bull. mar. biol. Lab. Woods Hole* **83,** 438-450.

98. BULLOCK, T. H., 1953. Predator recognition and escape responses of some intertidal gastropods in the presence of some starfish. *Behaviour* **5,** 130-140.

99. BULLOCK, T. H., 1965. Comparative aspects of superficial conduction systems in echinoids and asteroids. *Am. Zool.* **5,** 545-562.

100. BULLOCK, T. H. and NACHMANSOHN, D., 1942. Cholinesterase in primitive nervous systems. *J. cell. comp. Physiol.* **20,** 239-242.

101. BURNETT, A. L., 1960. The mechanism employed by the starfish *A. forbesi* to gain access to the interior of the bivalve *Venus mercenaria. Ecology* **41,** 583-584.

102. Burnett, A. L. and Anderson, J. M., 1955. The contractile properties of the retractor mechanism of the cardiac stomach in *Asterias forbesi. Anat. Rec.* **122,** 463-464.

103. Burrage, B. R., 1964. The possibility of paralytic effects of selected sea stars and brittle stars. *Trans. Kan. Acad. Sci.* **67,** 496.

104. Burton, M. P. M., 1964. Haemal system of regular echinoids. *Nature, Lond.* **204,** 1218.

105. Burton, M. P. M., 1966. Echinoid coelomic cells. *Nature, Lond.* **211,** 1095-1096.

106. Cannan, R. K., 1927. Echinochrome. *Biochem. J.* **21,** 184-189.

107. Carpenter, W. B., 1876. On the structure, physiology and development of *Antedon* (*Comatula*) *rosaceous. Proc. R. Soc.* **24,** 211-231 and 451-455.

108. Chadwick, H. C., 1893. Note on the water vascular system and haemal system of the Asteroidea. *Proc. Trans. Liverpool Biol. Soc.* **7,** 231-244.

109. Chadwick, H. C., 1923. Asterias. *L.M.B.C. Mem.* **25,**

110. Chaet, A. B., 1962. A toxin in the coelomic fluid of the scalded starfish *Asterias forbesi. Proc. Soc. exp. Biol. Med.* **109,** 791-794.

111. Chaet, A. B., 1964a. The shedding substance activity of starfish nerves. *Tex. Rep. Biol. Med.* **22,** 204.

112. Chaet, A. B., 1964b. A mechanism for obtaining mature gametes from starfish. *Biol. Bull. mar. biol. Lab. Woods Hole* **126,** 8-13.

113. Chaet, A. B., 1964c. Shedding substance and "shedhibin" from the nerves of the starfish *Patiria miniata. Am. Zool.* **4,** 407.

114. Chaet, A. B., 1965. Invertebrate adhering surfaces: Secretions of the starfish *Asterias forbesi* and the coelenterate *Hydra piradi. Ann. N.Y. Acad. Sci.* **118,** 921-929.

115. Chaet, A. B., 1966a. The gamete shedding substances of starfishes. A physiological-biochemical study. *Am. Zool.* **6,** 263-271.

116. Chaet, A. B., 1966b. Neurochemical control of gamete release in starfish. *Biol. Bull. mar. biol. Lab. Woods Hole* **130,** 43-58.

117. Chaet, A. B., 1967. Gamete release and shedding substance of sea stars, in *Echinoderm Biology,* Ed. N. Millott, Symp. zool. Soc. Lond. No. 20, pp. 13-24.

118. Chaet, A. B., Andrews, P. M. and Smith, R. H., 1964. The shedding substance of starfish nerve—its function and micro-assay. *Fed. Proc.* **23,** 204.

119. Chaet, A. B. and Cohen, S. I., 1958. A source of the toxic factor in scalded starfish. *Biol. Bull. mar. biol. Lab. Woods Hole* **115,** 347.

120. Chaet, A. B. and McConnaughty, R. A., 1959. Physiologic activity of nerve extracts. *Biol. Bull. mar. biol. Lab. Woods Hole* **117,** 407-408.

121. Chaet, A. B. and Musick, R. S., Jr., 1960. A method for obtaining gametes from *Asterias forbesi. Biol. Bull. mar. biol. Lab. Woods Hole* **119,** 292.

122. Chaet, A. B. and Philpott, D. E., 1960. Secretory structures in the tube foot of starfish. *Biol. Bull. mar. biol. Lab. Woods Hole* **119,** 308.

123. Chaet, A. B. and Philpott, D. E., 1961. Formation and possible function of the "secretory packets" of the starfish tube foot. *Biol. Bull. mar. biol. Lab. Woods Hole* **121,** 373.

124. Chaet, A. B. and Philpott, D. E., 1964. A new subcellular particle secreted by the starfish. *J. Ultrastruct. Res.* **11,** 354-362.

125. CHAET, A. B. and ROSE, R. A., 1961. Further studies on the gamete "shedding substance" from radial nerves. *Biol. Bull. mar. biol. Lab. Woods Hole* **121,** 385-386.

126. CHAET, A. B., SELLERS, R. and KENNAN, D., 1960. Further characteristics of a burn toxin from the starfish *Asterias forbesi. Anat. Rec.* **138,** 340.

127. CHAET, A. B. and SMITH, R. H., 1962. Role of gamete shedding substance from the starfish nerves. *Am. Zool.* **2,** 511.

128. CHANG, C. W., 1964. Structure and properties of Spinochrome H. *Diss. Abstr.* **25,** 2753.

129. CHANLEY, J. D., KOHN, S. K., NIGRELLI, R. F. and SOBOTKA, H., 1955. Further chemical analysis of Holothurin, the saponin-like steroid from the sea-cucumber. *Zoologica, N.Y.* **40,** 99.

130. CHANLEY, J. D., LEDEEN, R., WAX, J., NIGRELLI, R. F. and SOBOTKA, H., 1959. Holothurin: I. The isolation, properties and sugar components of Holothurin A. *J. Am. chem. Soc.* **81,** 5180-5183.

131. CHANLEY, J. D., MEZZETTI, T. and SOBOTKA, H., 1966. The Holothurinogenins. *Tetrahedron* **22,** 1857-1884.

132. CHANLEY, J. D., PERLSTEIN, J., NIGRELLI, R. F. and SOBOTKA, H., 1960. Further studies on the structure of Holothurin. *Ann. N.Y. Acad. Sci.* **90** (Art. 9), 902-905.

133. CHAPEAU, M., 1893. Sur la nutrition des Echinodermes. *Bull. Acad. r. Med. Belg.* **26,** (Ser. 3), 227-232.

134. CHERBONNIER, G., 1960. Sur la presence en Mer Noire de *Stereoderma kirschbergi* (Heller). *Hidrobiol. (Istanbul)* **5B,** 52-53.

134b. CHESHER, R. H., 1969. Destruction of Pacific Corals by the sea star *Acanthaster planci. Science, N.Y.* **165,** 280-283.

135. CHOE, S., 1962. Biology of the Japanese sea cucumber *Stichopus japonicus* Selenka. Fisheries Coll., Pusan National Univ. Pusan, Korea, pp. 226.

136. CHRISTENSEN, A. M., 1957. The feeding behaviour of the sea star *Evasterias troschelli* Stimpson. *Limnol. Oceanogr.* **2,** 180-197.

137. CLARK, A. M., 1962. *Starfishes and their relations.* Brit. Mus. Nat. Hist., London, pp. 1-119.

138. CLARK, F. W. and WHEELER, W. C., 1922. The inorganic constituents of marine invertebrates. *U.S. Geol. Surv. Prof. Papers* No. 124, 1-62.

139. CLARK, H. L., 1899. *Synapta vivipara*: A contribution to the morphology of echinoderms. *Mem. Boston Soc. Nat. Hist.* **5,** 53-88.

140. CLARK, H. L., 1915. The comatulids of the Torres Straits, with special reference to their habits and reactions. *Pap. Tortugas Lab.* **8,** 99-125.

141. CLARK, H. L., 1917. The habits and reactions of a comatulid *Tropimetra carinata. Pap. Tortugas Lab.* **11,** 111-119.

142. COBB, J. L. S., 1967. The innervation of the ampulla of the tube foot in the starfish *Astropecten irregularis. Proc. R. Soc.* B. **168,** 91-99.

143. COBB, J. L. S., 1968. Observations on the electrical activity within the retractor muscles of the lantern of *Echinus esculentus* using extracellular recording electrodes. *Comp. Biochem. Physiol.* **24,** 311-315.

144. COBB, J. L. S. and LAVERACK, M., 1966a. The lantern of *Echinus esculentus.* I. Gross anatomy and physiology. *Proc. R. Soc.* B. **164,** 624-640.

145. COBB, J. L. S. and LAVERACK, M., 1966b. The lantern of *Echinus esculentus.*

II. Fine structure of hyponeural tissue and its connections. *Proc. R. Soc.* B. **164**, 641-650.

146. Cobb, J. L. S. and Laverack, M., 1966c. The lantern of *Echinus esculentus*. III. The fine structure of the lantern retractor muscle and its innervation. *Proc. R. Soc.* B. **164**, 651-658.

147. Cobb, J. L. S. and Laverack, M., 1967. Neuromuscular systems in Echinoderms, in *Echinoderm Biology*, Ed. N. Millott, Symp. zool. Soc. Lond. No. 20, pp. 25-52.

148. Cocanour, B. and Allen, K., 1967. The breeding cycles of a sand dollar and a sea urchin. *Comp. Biochem. Physiol.* **20**, 327-331.

149. Coe, W. R., 1912. Echinoderms of Connecticut. *Bull. Conn. geol. nat. Hist. Surv.* **4**, 1-152.

150. Cohnheim, O., 1901. Versuche uber Resorption, Verdawing, und Stoffwechsel von Echindermen. *Hoppe-Seyler's Z. physiol. Chem.* **33**, 9-54.

151. Cole, L. J., 1913. Direction of locomotion of *Asterias*. *J. exp. Zool.* **14**, 1-32.

152. Cole, W. H., 1940. Composition of fluids and sera of some marine animals and of the sea water in which they live. *J. gen. Physiol.* **23**, 575-584.

153. Colwin, L. H., 1948. Note on the spawning of the holothurian *Thyone briareus* (Leseur). *Biol. Bull. mar. biol. Lab. Woods Hole* **95**, 296-306.

154. Cornil, L., Mosinger, M. and Calen, J., 1935a. La désintégration physiologique de l'apparail pigmentaire chez les holothuries. *C. r. Séanc. Soc. Biol.* **119**, 106-107.

155. Cornil, L., Mosinger, M. and Calen, J., 1935b. Sur la disposition reticulée du systéme pigmentaire chez les holothuries (*H. tubulosa*, *H. nigra*) et sa reactivité. *C. r. Séanc. Soc. Biol.* **118**, 1339-1341.

156. Cornman, I., 1941. Sperm activation by *Arbacia* egg extracts, with special reference to echinochrome. *Biol. Bull. mar. biol. Lab. Woods Hole* **80**, 202-207.

157. Cornman, I., 1963. Toxic properties of the saliva of *Cassis*. *Nature, Lond.* **200**, 88-89.

158. Costello, D. P., 1946. The swimming of *Leptosynapta*. *Biol. Bull. mar. biol. Lab. Woods Hole* **90**, 93-96.

159. Cottrell, G. A., 1967. Occurrence of dopamine and noradrenaline in the nervous tissue of some invertebrate species. *Br. J. Pharmac. Chemother.* **29**, 63-69.

160. Cousteau, J.-Y., 1963. *The Living Sea*. Hamish Hamilton, London. 212 pp.

161. Cowles, R. P., 1909a. Preliminary report on the behaviour of echinoderms. *Yearbook Carnegie Inst.* **8**, 128-129.

162. Cowles, R. P., 1909b. The movement of the starfish *Echinaster* towards light. *Zool. Anz.* **35**, 193-195.

163. Cowles, R. P., 1910. Stimuli produced by light and by contact with solid walls as factors in the behaviour of ophiuroids. *J. exp. Zool.* **9**, 387-416.

164. Cowles, R. P., 1911. Reaction to light and other points in the behaviour of the starfish. *Pap. Tortugas Lab.* **3**, 97-110.

165. Cowles, R. P., 1914. The influence of white and black walls on the direction of locomotion of the starfish. *J. Anim. Behav.* **4**, 380-382.

166. Creange, J. E. and Szego, C. M., 1967. Sulphation as a metabolic pathway for oestradiol in the sea urchin *Strongylocentrotus franciscanus*. *Biochem. J.* **102**, 898-904.

167. CRESCITELLI, F., 1945. A note on the absorption spectra of *Eudistylia gigantea* and of the pigment in the red corpuscles of *Cucumaria miniata* and *Molpadia intermedia*. *Biol. Bull. mar. biol. Lab. Woods Hole* **88,** 30-36.
168. CROZIER, W. J., 1915. The sensory reactions of *Holothuria surinamensis*. *Zool. Jb. Abt. Allg. Zool.* **35,** 233-297.
169. CROZIER, W. J., 1916. The rhythmic pulsations of the cloaca of holothurians. *J. exp. Zool.* **20,** 297-356.
170. CROZIER, W. J., 1918. The amount of bottom material ingested by holothurians (*Stichopus*). *J. exp. Zool.* **26,** 379-389.
171. CROZIER, W. J., 1920. The volume of water involved in the cloacal pumping of holothurians (*Stichopus*). *Biol. Bull. mar. biol. Lab. Woods Hole* **39,** 130-132.
172. CROZIER, W. J., 1935a. The geotropic response in *Asterina*. *J. gen. Physiol.* **18,** 729-737.
173. CROZIER, W. J., 1935b. On reversal of geotropism in *Asterina*. *J. gen. Physiol.* **18,** 739-742.
174. CUENOT, L., 1890. Sur le systéme madreporique des Echinodérmes. *Zool. Anz.* **13,** 315-318.
175. CUENOT, L., 1891a. Études morphologiques sur les Echinodérmes. *Archs. Biol.* **11,** 312-680.
176. CUENOT, L., 1891b. Études sur le sang et les glandes lymphatiques dans la serie animale. II. Invertébrès. *Archs. Zool. exp. gen.* **9,** 595-670.
177. CUENOT, L., 1901. Etudes physiologiques sur les Asteriès. *Archs. Zool. exp. gen.* **9,** 233-259.
178. CUENOT, L., 1906. Rôle biologique de la coagulation du liquides coelomiques des Oursins. *C. r. Séanc. Soc. Biol.* **61,** 255-256.
179. CUENOT, L., 1948. Anatomie, Ethologie et Systematique des Echinodermes, in *Traite de Zoologie*, Ed. P. Grasse. Masson. Paris.
180. CUTRESS, B. M., 1965. Observations on growth in *Eucidaris tribuloides* (Lamarck) with special reference to the origin of the oral primary spines. *Bull. mar. Sci. Gulf Caribb.* **15,** 797-834.
181. DABROWA, N., LANDAU, J. W., NEWCOMER, V. D. and PLUNKETT, O. A., 1964. A survey of the tide washed coastal areas of Southern California for fungi potentially pathogenic to man. *Mycopath. Mycol. appl.* **24,** 137-150.
182. D'AGOSTINO, A. S. and FARMANFARMAIAN, A., 1960. Transport of nutrients in the holothurian *Leptosynapta inhaerens*. *Biol. Bull. mar. biol. Lab. Woods Hole* **119,** 301.
183. DAHLGREN, U., 1916. The production of light by animals. *J. Frankl. Inst.* **181,** 377-400.
184. DAKIN, W. J., 1908. Variations in the osmotic concentrations of the blood and coelomic fluids of aquatic animals caused by changes in the external medium. *Biochem. J.* **3,** 473-490.
185. DAN, K. and DAN, J. C., 1941. Spawning habit of the crinoid *Comanthus japonicus*. *Jap. J. Zool.* **9,** 555-564.
186. DAN, K. and KUBOTA, H., 1960. Data on the spawning of *Comanthus japonicus* between 1937 and 1950. *Embryologia* **5,** 21-37.
187. DAVENPORT, D., 1950. Studies in the physiology of commencalism. *Biol. Bull. mar. biol. Lab. Woods Hole* **98,** 81-93.
188. DAVENPORT, D. and HICKOK, J. F., 1951. Studies in the physiology of

commencalism. 2. The polynoid genus *Arctonoe* and *Halosydna*. *Biol. Bull. mar. biol. Lab. Woods Hole* **100,** 71-83.

189. Davidson, E., 1953. Clotting of the perivisceral fluid of the sand dollar *Echinarachnius parma*. *Biol. Bull. mar. biol. Lab. Woods Hole* **105,** 372.

190. Davis, W. P., 1966. Observations on the biology of the ophiuroid *Astrophyton muricatum*. *Bull. mar. Sci. Gulf Caribb.* **16,** 435-444.

191. Defretin, R., 1952. Sur les mucocytes des podia de quelques Echinodérmes. *C. r. hebd. Séanc. Acad. Sci. Paris* **234,** 1806.

192. Delage, Y., 1902. Effets de l'excision du madreporite chez les astèries. *C. r. hebd. Séanc. Acad. Sci. Paris* **135,** 841-842.

193. Delage, Y., 1903. Sur la non-regeneration des sphèridies chez les Oursins. *C. r. hebd. Séanc. Acad. Sci. Paris* **137,** 681-682.

194. Delaunay, H., 1931. L'Excrétion azotée des Invertebrès. *Biol. Rev.* **6,** 265-301.

195. Delaunay, H., 1934. Le metabolisme de l'ammoniaque d'après les recherches relatives aux Invertebrès. *Ann. Physiol. physico-chim. Biol.* **10,** 695-725.

196. Demoor, J. and Chapeaux, M., 1892. Contribution a la physiologie nerveuse des Echinodermes. *Tijdschr. ned. dierk. Vereen.* **3** (Ser. 2.), 108-168.

197. DesArtes, L., 1910. Uber die Lebensweise von *Amphiura chiajei* unter Berushsichtigung der Anatomischen Verhaltnisse. *Bergens Mus. Arb. Natur. Rekke.* **12,** 1-10.

198. D'Iakonov, A. M., 1955. On the possibility of the existence of echinoderms in oceans of lowered salinity. *C. r. Acad. Sci. Moscow* **102,** 373-374.

199. Diebschlag, E., 1938. Ganzheitliches verhalten und lernen bei Echinodermen. *Z. vergl. Physiol.* **25,** 612-654.

200. Dimelow, E. J., 1958. Pigments present in arms and pinnules of the crinoid *Antedon bifida*. *Nature, Lond.* **182,** 812.

201. Donahue, J. K., 1940. Occurrence of estrogens in the ovaries of certain invertebrates. *Endocrinology* **27,** 149-152.

202. Donnellon, J. A., 1938. An experimental study of clot formation in the perivisceral fluid of *Arbacia*. *Physiol. Zool.* **11,** 389-397.

203. Doree, C., 1909. The occurrence and distribution of cholesterol and allied bodies in the animal kingdom. *Biochem. J.* **4,** 72-106.

204. Doyle, W. L., 1964. Basement membrane and its precursor in haemal vessels of a holothurian. *J. Cell. Biol.* **23,** 26A.

205. Doyle, W. L., 1967. Vesiculated axons in haemal vessels of a holothurian, *Cucumaria frondosa*. *Biol. Bull. mar. biol. Lab. Woods Hole* **132,** 329-336.

206. Doyle, W. L. and McNeill, G. F., 1964. The fine structure of the respiratory tree in *Cucumaria*. *Q. Jl Microsc. Sci.* **105,** 7-11.

207. Drumm, P. J., O'Connor, W. F. and Renouf, L. P., 1946. The pigments of sponges. I. The lipid pigments of the sponge *Hymeniacidon sanguineum*. *Biochem. J.* **39,** 208-210.

208. Dubois, R., 1914. Action de la lumière sur les echinodermes. *C. r. Congr. int. Zool.* **9,** 148.

209. DuBuy, H. G., 1936a. The physiology of an invertebrate smooth muscle (Retractor of *Thyone briareus*). *Am. J. Physiol.* **116,** 22-23.

210. DuBuy, H. G., 1936b. Separation of the conducting and contractile elements in the retractor muscle of *Thyone briareus*. *Biol. Bull. mar. biol. Lab. Woods Hole* **71,** 408-409.

211. DUGAL, L. P., 1939. The use of calcareous shell to buffer the product of anaerobic glycollysis in *Venus mercenaria*. *J. cell. comp. Physiol.* **13,** 235-251.

212a. DURHAM, H. E., 1888. The emigration of corpuscles in the starfish. *Proc. R. Soc.* **43,** 327.

212b. DURHAM, H. E., 1891. On wandering cells in echinoderms and more especially with regard to excretory function. *Q. Jl Microsc. Sci.* **33,** 81.

213. DUVAL, M. and PORTIER, P., 1927. Sur la teneur en gaz carbonique total du sang des Invertebrès d'eau douce et des Invertebrès marins. *C. r. hebd. Séanc. Acad. Sci. Paris* **184,** 1594-1596.

214. EAKIN, R. M. and WESTFALL, J. A., 1964. Further observations on the fine structure of some invertebrate eyes. *Z. Zellforsch. mikrosk. Anat.* **62,** 310-332.

215. EBERT, T. A., 1967. Negative growth and longevity in the purple sea urchin *Strongylocentrotus purpuratus*. *Science, N.Y.* **157,** 557-558.

216. EICHBERG, J., GILBERTSON, J. R. and KARNOVSKY, M. L., 1961. Neutral plasmologens analogous to the neutral triglycerides. *J. biol. Chem.* **236,** 15-16 pc.

217. ENDEAN, R., 1958. The coelomocytes of *H. leucospilotica*. *Q. Jl Microsc. Sci.* **99,** 47-60.

218. ENNOR, A. H. and MORRISON, J. F., 1958. Biochemistry of the phosphagens and related guanidines. *Physiol. Rev.* **38,** 631-674.

219. ENNOR, A. H. and ROSENBERG, H., 1962. Lombricine and serine ethanolamine phosphodiester, in *Amino Acid Pools*, Ed. J. T. Holden, Elsevier, Amsterdam. pp. 187-193.

220. ENRIQUES, P., 1903. Digestione, circolazione e assorbimento nelle Oloturie. *Archo. zool. ital.* **1,** 1-58.

221. EPPLEY, R. W. and LASKER, R., 1959. Alginase in the sea urchin *Strongylocentrotus purpuratus*. *Science, N.Y.* **129,** 214-215.

222. VON EULER, U. S., CHAVES, N. and TEODOSIO, N., 1952. Effect of acetylcholine, noradrenaline, adrenaline and histamine on isolated organs of *Aplysia* and *Holothuria*. *Acta Physiol. Latinoamer.* **2,** 101-106.

223. VON EULER, U. S. and HELLSTROM, H., 1934. Uber Asterinsaure, eine Carotinoidsaure aus Seesternen. *Hoppe-Seyler's Z. physiol. Chem.* **223,** 97-98.

224. EVANS, A. C., 1948. Some effects of earthworms on soil structure. *Ann. appl. biol.* **35,** 1-13.

225. EWER, D. W., 1965. Networks and spontaneous activity in echinoderms and platyhelminthes. *Am. Zool.* **5,** 563-572.

226. FAGERLUND, U. H. M. and IDLER, D. R., 1960. Marine sterols. VI. Sterol biosynthesis in molluscs and echinoderms. *Can. J. Biochem. Physiol.* **38,** 997-1002.

227. FANGE, R. and OSTLUND, E., 1954. The effects of adrenaline, noradrenaline, tyramine and other drugs on the isolated heart from marine vertebrates and a cephalopod. *Acta Zool. Stockh.* **35,** 289-305.

228. FARMANFARMAIAN, A., 1959. The respiratory surface of the purple sea urchin. *Anat. Rec.* **134,** 561.

229. FARMANFARMAIAN, A., 1966. The respiratory physiology of echinoderms, in *Physiology of Echinodermata*. Ed. R. A. Boolotian. J. Wiley, New York, pp. 245-265.

230. FARMANFARMAIAN, A. and GIESE, A. C., 1963. Thermal tolerance and

acclimatization in the western purple sea urchin, *Strongylocentrotus purpuratus. Physiol. zool.* **36,** 237-243.

231. Farmanfarmaian, A., Giese, A. C., Boolootian, R. A. and Bennett, J., 1958. Annual reproductive cycles in four species of west coast starfishes. *J. exp. Zool.* **138,** 355-367.

232. Farmanfarmaian, A. and Phillips, J. H., 1962. Digestion storage and translocation of nutrients in the purple sea urchin, *Strongylocentrotus purpuratus. Biol. Bull. mar. biol. Lab. Woods Hole* **123,** 105-120.

233. Fechter, H., 1965. Uber die funktion der Madreporenplatte der Echinoidea. *Z. vergl. Physiol.* **51,** 227-257.

234. Feder, H. M., 1955. On the methods used by the starfish *Pisaster ochraceous* in opening three types of bivalve mollusc. *Ecology* **36,** 764-766.

234b. Feder, H. M., 1959. The food of the starfish *Pisaster ochraceous* along the California coast. *Ecology* **40,** 721-724.

235. Feder, H. M., 1963. Gastropod defensive responses and their effectiveness in reducing predation by starfishes. *Ecology* **44,** 505-512.

236. Feder, H. M. and Lasker, R., 1964. Partial purification of a substance from starfish tube feet which elicits escape responses in gastropod molluscs. *Life Sci.* **3,** 1047-1051.

237. Feigen, G. A., Santz, E. and Allender, C. B., 1966. Studies on the mode of action of sea urchin toxin. I. Conditions affecting release of histamine and other agents from isolated tissues. *Toxicon* **4,** 161-175.

238. Fell, H. B., 1965. The early evolution of the Echinozoa. *Breviora,* No. 219, 1-17.

239. Fenchel, T., 1965. Feeding biology of the sea star *Luidia sarsi* Duben and Koren. *Ophelia* **2,** 223-236.

240a. Ferguson, J. C., 1963. An autoradiographic study of the distribution of ingested nutrients in the starfish *Asterias forbesi. Am. Zool.* **3,** 524.

240b. Ferguson, J. C., 1964a. Nutrient transport in starfish. I. Properties of the coelomic fluid. *Biol. Bull. mar. biol. Lab. Woods Hole* **126,** 33-53.

241. Ferguson, J. C., 1964b. Nutrient transport in starfish. II. Uptake, of nutrients by isolated organs. *Biol. Bull. mar. biol. Lab. Woods Hole* **126,** 391-406.

242. Ferguson, J. C., 1966. Mechanical responses of isolated starfish digestive glands to metabolic inhibitors drugs, and nutrients. *Comp. Biochem. Physiol.* **19,** 259-266.

243. Ferguson, J. C., 1967. Utilization of dissolved exogenous nutrients by the starfishes *Asterias forbesi* and *Henricia sanguinolenta. Biol. Bull. mar. biol. Lab. Woods Hole* **132,** 161-173.

244. Fish, J. D., 1967a. The digestive system of the holothurian *Cucumaria elongata.* I. Structure of the gut and haemal system. *Biol. Bull. mar. biol. Lab. Woods Hole* **132,** 337-353.

245. Fish, J. D., 1967b. The digestive system of the holothurian *Cucumaria elongata.* II. Distribution of the digestive enzymes. *Biol. Bull. mar. biol. Lab. Woods Hole* **132,** 354-361.

246. Fish, J. D., 1967c. The biology of *Cucumaria elongata. J. mar. biol. Ass. U.K.* **47,** 129-143.

247. Florey, E., 1956. The action of Factor I on certain invertebrate organs. *Can. J. Biochem. Physiol.* **34,** 669-681.

248. FLOREY, E., 1962. Comparative neurochemistry: inorganic ions, amino acids and possible transmitter substances of Invertebrates, in *Neurochemistry*. Ed. Elliott, K. A. C., Page, I. H. and Quastel, J. H. Thomas, Springfield, U.S.A., p. 1035.

249. FLOREY, E. and MCLENNAN, H., 1959. The effects of Factor I and GABA on smooth muscle preparations. *J. Physiol. Lond.* **145,** 66-67.

250. FLORKIN, M., 1962a. La regulation isoosmotiques intracellulaire chez les Invertebrès marins euryhalins. *Bull. Acad. Belg. Cl. Sci.* **48,** 687-694.

251. FLORKIN, M., 1962b. Regulation anisosmotiques extracellulaire, regulation isoosmotiques intracellulaire et euryhalinite. *Ann. Soc. Zool. Belg.* **92,** 183-186.

252. FLORKIN, M. and DUCHATEAU, G., 1943. Les formes du systeme enzymatique de l'uricolyse et l'evolution du catabolisme purique chez les animaux. *Archs. int. Physiol.* **53,** 267.

253. FOETTINGER, A., 1880. Sur l'existence de l'haemoglobine chez les echinodermes. *Archs. Biol. Paris* **1,** 405-413.

254. FONTAINE, A. R., 1955. Secretion of a highly sulphated acid mucopolysaccharide by the brittle star *Ophiocomina nigra. Nature, Lond.* **176,** 606-607.

255. FONTAINE, A. R., 1962a. The colours of *Ophiocomina nigra* (Abildgaard). I. Colour variation and its relation to distribution. *J. mar. biol. Ass. U.K.* **42,** 1-8.

256. FONTAINE, A. R., 1962b. The colours of *Ophiocomina nigra* (Abildgaard). II. The occurrence of melanin and fluorescent pigments. *J. mar. biol. Ass. U.K.* **42,** 9-31.

257. FONTAINE, A. R., 1962c. The colours of *Ophiocomina nigra* (Abildgaard). III. Carotenoid pigments. *J. mar. biol. Ass. U.K.* **42,** 33-47.

258. FONTAINE, A. R., 1962d. Neurosecretion in the ophiuroid *Ophiopholis aculeata. Science, N.Y.* **138,** 908-909.

259. FONTAINE, A. R., 1964. The integumentary mucus secretions of the ophiuroid *Ophiocomina nigra. J. mar. biol. Ass. U.K.* **44,** 145-162.

260. FORSTER, G. R., 1959. The ecology of *Echinus esculentus* L. Quantitative distribution and rate of feeding. *J. mar. biol. Ass. U.K.* **38,** 361-367.

261. FOSSE, R. and BRUNEL, A., 1929. Distribution of Allantoinase. *C. r. hebd. Séanc. Acad. Sci. Paris* **188,** 1067.

262. FOX, D. L., 1953. *Animal Biochromes and Structural Colours.* Camb. Univ. Press, London, 378 pp.

263. FOX, D. L. and SCHEER, B. T., 1941. Comparative studies of the pigments of some Pacific coast echinoderms. *Biol. Bull. mar. biol. Lab. Woods Hole* **80,** 441-455.

264. FOX, H. M., 1922. Lunar periodicity in reproduction. *Nature, Lond.* **109,** 237-238.

265. FOX, H. M., 1924a. Lunar periodicity in reproduction. *Proc. R. Soc.* B. **95,** 523-550.

266. FOX, H. M., 1924b. The spawning of echinoids. *Proc. Camb. phil. Soc. biol. Sci.* **1,** 71-74.

267. FRAENKEL, G., 1928. Uber den auslosungsriez des Umdrehreflexes bei seesternen und schlangensternen. *Z. vergl. Physiol.* **7,** 165-378.

268. FRANSSEN, J. and JEUNIAUX, C., 1965. Digestion de l'acide alginique chez les invertebrès. *Cah. Biol. mar.* **6,** 1-21.

269. FREDERICQ, L., 1901. Sur la concentration moleculaire du sang et des tissus chez les animaux aquatiques. *Bull. Acad. Belg. Cl. Sci.* **8,** 428-454.

270. FREEMAN, P. J., 1966. Observations on osmotic relationships in the holothurian *Opheodesoma spectabilis. Pacific Sci.* **20,** 60-69.

271. FREEMAN, W. P. and SIMON, S. E., 1964 .The histology of holothuroidean muscle. *J. cell. comp. Physiol.* **63,** 25-38.

272. FRIEDHEIM, E. A.-H., 1932. Sur deux ferments respiratoire accessoires d'origine animale. *C. r. Séanc. Soc Biol.* **111,** 505-507.

273. FRIES, S. L. and DURANT, R. C., 1963. Protective interactions of physostigmine and Holothurin A at the mammalian neuromuscular junction. *Biochem. Pharmac.* **12,** 166.

274. FRIES, S. L. and DURANT, R. C., 1965. Blockage phenomena at the mammalian neuromuscular junction—competition between reversible anticholinesterases and an irreversible toxin. *Toxicol. Appl. Pharmac.* **7,** 373-381.

275. FRIES, S. L., DURANT, R. C., CHANLEY, J. D. and MEZZETTI, T., 1965. Some structural requirements underlying Holothurin A interactions with synaptic chemoreceptors. *Biochem. Pharmac.* **14,** 1237-1247.

276. FRIES, S. L., STANDAERT, F. G., WHITCOMB, E. R., NIGRELLI, R. F., CHANLEY, J. D. and SOBOTKA, H., 1959. Some pharmacologic properties of Holothurin, an active neurotoxin from the sea cucumber. *J. Pharmac. exp. Ther.* **126,** 323-329.

277. FRIES, S. L., STANDAERT, F. G., WHITCOMB, E. R., NIGRELLI, R. F., CHANLEY, J. D. and SOBOTKA, H., 1960. Some pharmacologic properties of Holothurin A, a glycosidic mixture from the sea cucumber. *Ann. N.Y. Acad. Sci.* **90** (Art. 3), 893-901.

278. FUJI, A., 1960a. Studies on the biology of the sea urchin. I. Superficial and histological gonadal changes in gametogenic processes of two sea urchins, *Strongylocentrotus nudus* and *S. intermedius. Bull. Fac. Fish. Hokkaido Univ.* **11,** 1-14.

279. FUJI, A., 1960b. Studies on the biology of the sea urchin. II. Size at first maturity and sexuality of two sea urchins. *Bull. Fac. Fish. Hokkaido Univ.* **11,** 43-48.

280. FUJI, A., 1960c. Studies on the biology of the sea urchin. III. Reproductive cycle of two sea urchins *Strongylocentrotus nudus* and *S. intermedius. Bull. Fac. Fish. Hokkaido Univ.* **11,** 49-57.

281. FUJI, A., 1960d. Studies on the biology of the sea urchin. IV. Histological observation of the food canal of *Strongylocentrotus intermedius. Bull. Fac. Fish. Hokkaido Univ.* **11,** 195-202.

282. FUJI, A., 1962. Studies on the biology of the sea urchin. V. Food consumption of *Strongylocentrotus intermedius. Jap. J. Ecol.* **12,** 181-186.

283. FUJIWARA, T., 1935. On the poisonous pedicellaria of *Toxopneustes pileolus. Annotnes zool. jap.* **15,** 62-69.

284. GALTSOFF, P. S. and LOOSANOFF, V. L., 1939. Natural history and methods of controlling the sea star *Asterias forbesi. Bull. U.S. Bur. Fish.* **49,** 75-132.

285. GARREY, W. E., 1904. Osmotic pressure of sea water and of the blood of marine animals. *Biol. Bull mar. biol. Lab. Woods Hole* **8,** 257-270.

286. GAY, W. S. and SIMON, S. E., 1964. Metabolic control in holothuroidean muscle. *Comp. Biochem. Physiol.* **11,** 183-192.

287. Geddes, P., 1880. On the coalescence of amoeboid cells into plasmodia and on the so-called coagulation of invertebrate fluids. *Proc. R. Soc.* **30,** 252-255.

288. Gellhorn, E., 1927. Vergleichend-physiologische Untersuchungen uber die Pufferungspotenz von Blüt und Körpersöften. *Pflügers Arch. ges. Physiol.* **216,** 253-266.

289. Gemmill, J. F., 1912. The locomotor function of the lantern in *Echinus* with observations on other allied lantern activities. *Proc. R. Soc.* B. **85,** 84-109.

290. Gemmill, J. F., 1914. The development and certain points in the adult structure of *Asterias rubens* L. *Phil. Trans. R. Soc.* B. **205,** 213-294.

291. Gemmill, J. F., 1915. On the ciliation of asteroids and on the question of ciliary nutrition in certain species. *Proc. zool. Soc. Lond.* (1915), pp. 1-19.

292. Gezelius, G., 1963. Adaptation of the sea urchin *Psammechinus miliaris* to different salinities. *Zool. Bidr. Upps.* **35,** 329-337.

293. Ghiradella, H. T., 1965. The reaction of two starfishes *Patiria miniata* and *Asterias forbesi* to foreign tissue in the coelom. *Biol. Bull. mar. biol. Lab. Woods Hole* **128,** 77-89.

294. Giese, A. C., 1959. Comparative physiology: annual reproductive cycles of marine invertebrates. *Ann. Rev. Physiol.* **21,** 547-576.

295. Giese, A. C., 1961. Further studies on *Allocentrotus fragilis,* a deep sea echinoid. *Biol. Bull. mar. biol. Lab. Woods Hole* **121,** 141-150.

296. Giese, A. C., 1966. On the biochemical constitution of some echinoderms, in *Physiology of Echinodermata.* John Wiley, New York, pp. 757-796.

297. Giese, A. C., 1967. Changes in body component indexes and respiration with size, in the purple sea urchin *Strongylocentrotus purpuratus. Physiol. Zool.* **40,** 194-200.

298. Giese, A. C. and Farmanfarmaian, A., 1963. Resistance of the purple sea urchin to osmotic stress. *Biol. Bull. mar. biol. Lab. Woods Hole* **124,** 182-192.

299. Giese, A. C., Farmanfarmaian, A., Hilden, S. and Doezema, P., 1966. Respiration during the reproductive cycle in the sea urchin *Strongylocentrotus purpuratus. Biol. Bull. mar. biol. Lab. Woods Hole* **130,** 192-201.

300. Giese, A. C., Greenfield, L., Huang, H., Farmanfarmaian, A., Boolootian, R. and Lasker, R., 1959. Organic productivity in the reproductive cycle of the purple sea urchin. *Biol. Bull. mar. biol. Lab. Woods Hole* **116,** 49-58.

301. Giese, A. C., Krishnaswamy, S., Vasu, B. S. and Lawrence, J., 1964. Reproductive and biochemical studies on a sea urchin *Stomopneustes variolaris* from Madras harbour. *Comp. Biochem. Physiol.* **13,** 367-380.

302. Giordano, M. F., Harper, H. A. and Filice, F. P., 1950. The amino acids of a starfish and a sea urchin. *Wasmann J. Biol.* **8,** 129-132.

303. Gislén, T., 1924. Echinoderm studies. *Zool. Bidr. Upps.* **9,** 1-316.

304. Glaser, O., 1907. Movement and problem solving in *Ophiura. J. exp. Zool.* **4,** 205-220.

305. Glaser, R. and Lederer, E., 1939. Echinochrome et Spinochrome: dérivés méthylés distribution pigments associés. *C. r. hebd. Séanc. Acad. Sci. Paris* **208,** 1939.

306. Glynn, P. W., 1965. Active movements and other aspects of the biology

of *Astichopus* and *Leptosynapta*. *Biol. Bull. mar. biol. Lab. Woods Hole* **129,** 106-127.

307. GOODBODY, I., 1960. The feeding mechanism in the sand dollar *Mellita sexiesperforata* (Leske). *Biol. Bull. mar. biol. Lab. Woods Hole* **119,** 80-86.

308. GOODWIN, T. W. and FOX, D. L., 1955. Some observations on pigments of the Pacific sand dollars, *Dendraster excentricus* and *D. laevis*. *Experientia* **11,** 270-271.

309. GOODWIN, T. W., LEDERER, E. and MUSAJO, L., 1951. The nomenclature of the spinochromes of sea urchins. *Experientia* **7,** 375-376.

310. GOODWIN, T. W. and SRISUKH, S., 1950. A study of the pigments of the sea urchins *Echinus esculentus* and *Paracentrotus lividus*. *Biochem. J.* **47,** 69-76.

311. GOUGH, J. H. and SUTHERLAND, M. D., 1967. Marine pigments. VII. 3-acetyl 2.5.6.7. tetrahydroxy 1 : 4 naphthoquinone, a new spinochrome from *Salmacis sphaeroides*. *Aust. J. Chem.* **20,** 1693-1697.

312. GRABER, V., 1889. Uber die Empfindlichkeit einiger Meertiere gigen Riechstoffe. *Biol. Zbl.* **8,** 743-754.

313. GREENFIELD, L., GIESE, A. C., FARMANFARMAIAN, A. and BOOLOOTIAN, R. A., 1958. Cyclical biochemical changes in several echinoderms. *J. exp. Zool.* **139,** 507-524.

314. GRIFFITHS, A. B., 1888. Further researches on the physiology of invertebrates. *Proc. R. Soc.* **44,** 325-328.

315. GRIFFITHS, M., 1965. A study of the synthesis of naphthoquinone pigments by the larvae of two species of sea urchin and their reciprocal hybrids. *Devl. Biol.* **11,** 433-447.

316. HAGERMAN, D. D., WELLINGTON, F. M. and VILLEE, C. A., 1957. Estrogens in marine invertebrates. *Biol. Bull. mar. biol. Lab. Woods Hole* **112,** 180-183.

317. HAMANN, O., 1887. On the phylogeny and anatomy of the Echinodermata. *Ann. Mag. nat. Hist.* **20,** 361-378.

318. HANCOCK, D., 1955. The feeding behaviour of starfish on Essex oyster beds. *J. mar. biol. Ass. U.K.* **34,** 313-331.

319. HANCOCK, D., 1957. The feeding behaviour of *Psammechinus miliaris*. *Proc. zool. Soc. Lond.* **129,** 255-262.

320. HANCOCK, D., 1958. Notes on starfish on an Essex oyster bed. *J. mar. biol. Ass. U.K.* **37,** 565-589.

321. HANCOCK, D., 1965. Adductor muscle size in Danish and British mussels and its relation to starfish predation. *Ophelia* **2,** 253-267.

322. HARRIS, E. J. and SJODIN, R. A., 1961. Kinetics of exchange and net movement of frog muscle potassium. *J. Physiol. Lond.* **155,** 221-245.

323. HARRISON, G. and PHILPOTT, D., 1966. Subcellular particles in echinoderm tube feet. I. Class Asteroidea. *J. Ultrastruct. Res.* **16,** 537-547.

324. HARRISS, R. C. and PILKEY, O. H., 1966. Temperature and salinity control of the concentration of skeletal sodium, manganese and iron in *Dendraster excentricus*. *Pacific Sci.* **20,** 235-238.

325. HARTLINE, H. K., WAGNER, H. G. and MACNICHOL, E. F., 1952. The peripheral origin of nervous activity in the visual system. *Cold Spring Harb. Symp. quant. Biol.* **16,** 125-141.

326. HARTMAN, H. B. and CHAET, A. B., 1962. Gamete shedding with radial nerve extracts. *Fed. Proc.* **21,** 363.

327. HARTMANN, M., SCHARTAU, O., KUHN, R. and WALLENFELS, K., 1939. Ueber die Sexualstoffe der Seeigel. *Naturwissenschaften* **27,** 433.

328. HARTOG, M. M., 1887. The true nature of the madreporic system. *Ann. Mag. nat. Hist.* **20,** 321-326.

329. HARVEY, E. N., 1952. *Bioluminescence.* Academic Press, New York, 649 pp.

330. HARVEY, H. W., 1955. *The Chemistry and Fertility of Seawaters.* Cambridge Univ. Press, London, 224 pp.

331. HASHIMOTO, Y. and YASUMOTO, T., 1960. Confirmation of saponin as a toxic principle of starfish. *Bull. Jap. Soc. scient. Fish.* **26,** 1132-1138.

332. HATANAKA, M. and KOSAKA, M., 1959. Biological studies on the population of the starfish *Asterias amurensis* in Sandai Bay. *Tohoku J. agric. Res.* **9,** 159-178.

333. HATHAWAY, R. R., 1965. Conversion of estradiol 17β by sperm preparations of sea urchins and oysters. *Gen. comp. Endocr.* **5,** 504.

334. HAUROWITZ, F. and WAELSCH, H., 1926. Vergleichende chemische Untersuchungen an Holothurien und Aktinen. *Hoppe-Seyler's Z. physiol. Chem.* **161,** 318.

335. HAYASHI, R., 1935. Studies on the morphology of Japanese sea stars. I. Anatomy of *Henricia sanguinolenta. J. Fac. Sci. Hokkaido Imp. Univ.* Ser. 6, **4,** 1-26.

336. HAZELHOFF, E. H., 1938. Uber die Ausnutzung des Sauerstoffs bei Verscheidenen Wassertieren. *Z. vergl. Physiol.* **26,** 306-327.

337. HENRI, V., 1903a. Étude physiologiques des muscles longitudinaux chez le *Stichopus regalis. C. r. Séanc. Soc. Biol.* **55,** 1194-1195.

338. HENRI, V., 1903b. Études des contractions rhythmiques des Vaisseaux et du poumon aquaux chez les Holothuries. *C. r. Séanc. Soc. Biol.* **55,** 1314-1316.

339. HENRI, V. and KAYALOFF, E., 1906. Études des toxines contenues dans les pédicellaires des Oursins. *C. r. Séanc. Soc. Biol.* **60,** 884-886.

340. HENRI, V. and LALOU, S., 1903a. Regulation osmotiques des liquides internes chez les Echinodermes. *C. r. hebd. Séanc. Acad. Sci. Paris* **137,** 721.

341. HENRI, V. and LALOU, S., 1903b. Regulation osmotiques des liquides internes chez les Oursins. *C. r. Séanc. Soc. Biol.* **55,** 1242.

342. HENRI, V. and LALOU, S., 1903c. Regulation osmotiques des liquides internes chez les Holothuries. *C. r. Séanc. Soc. Biol.* **55,** 1243.

343. HENRI, V. and LALOU, S., 1904. Regulation osmotiques des liquides internes chez les Echinodermes. *J. Physiol. Path. gen.* **6,** 9.

344. HESS, C., 1915. Untersuchungen uber den Lichtsinn bei Echinodermen. *Arch. ges. Physiol.* **160,** 1-26.

345. HETZEL, H. R., 1963. Studies on holothurian coelomocytes. I. A survey of coelomocyte types. *Biol. Bull. mar. biol. Lab. Woods Hole* **125,** 289-301.

346. HETZEL, H. R., 1965. Studies on holothurian coelomocytes. II. Origin of coelomocytes and the formation of brown bodies. *Biol. Bull. mar. biol. Lab. Woods Hole* **128,** 102-111.

347. HIESTAND, W. A., 1940. Oxygen consumption of *Thyone briareus* (Holothuroidea) as a function of oxygen tension and hydrogen ion concentration of the surrounding medium. *Trans. Wis. Acad. Sci. Arts. Letts.* **32,** 167-175.

348. HIESTAND, W. A., 1943. Action of certain drugs on the sea star *Asterias forbesi. Proc. Soc. exp. Biol. Med.* **52,** 85.

349. Hill, A. V., 1926. The viscous elastic properties of smooth muscle. *Proc. R. Soc.* B. **100,** 108-115.

350. Hinard, G. and Fillon, R., 1921. Sur la composition chimique des *Asterias. C. r. hebd. Séanc. Acad. Sci. Paris* **173,** 935-937.

351. Hogben, L. and van der Lingen, J., 1928. On the occurrence of haemoglobin and of erythrocytes in the perivisceral fluid of a holothurian. *J. exp. Biol.* **5,** 292-294.

352. Holland, L. Z., Giese, A. C. and Phillips, J. H., 1967. Studies on the perivisceral coelomic fluid protein concentration during seasonal and nutritional changes in the purple sea urchin. *Comp. Biochem. Physiol.* **21,** 361-371.

353. Holland, N. D. and Nimitz, A., 1964. An autoradiographic and histochemical investigation of the gut mucopolysaccharides of the purple sea urchin (*S. purpuratus*). *Biol. Bull. mar. biol. Lab. Woods Hole* **127,** 280-293.

354. Holland, N. D., Phillips, J. H. and Giese, A. C., 1965. An autoradiographic investigation of coelomocyte production in the purple sea urchin (*S. purpuratus*). *Biol. Bull. mar. biol. Lab. Woods Hole* **128,** 259-270.

355. Holmes, S. J., 1912. Phototaxis in the sea urchin *Arbacia punctulata. J. Anim. Behav.* **2,** 126-136.

356. Hoshiai, T., 1963. Some observations on the swimming of *Labidoplax dubia. Bull. mar. biol. Sta. Asamushi* **11,** 167-170.

357. Howell, W. H., 1885. Note on the presence of haemoglobin in Echinoderms. *Johns Hopk. Univ. Stud. biol. Lab.* **3,** 289-291.

358. Howell, W. H., 1885b. Chemical composition and coagulation of the blood of *Limulus polyphemus*, *Callinectes hastatus* and *Cucumaria* sp. *Johns Hopk. Univ. Circ.* **5,** 5.

359. Huang, H. and Giese, A. C., 1958. Tests for digestion of algal polysaccharides by some marine herbivores. *Science, N.Y.* **127,** 475.

360. Hulings, N. C. and Hemlay, D. W., 1963. An investigation of the feeding habits of two species of sea stars. *Bull. mar. Sci. Gulf. Caribb.* **13,** 354-359.

361. Hyman, L., 1929. The effect of oxygen tension on oxygen consumption in some planarians and echinoderms. *Physiol. Zool.* **2,** 505-534.

362. Hyman, L., 1955. *The Invertebrates, IV. Echinodermata.* McGraw-Hill, New York, 763 pp.

363. Ikeda, H., 1941. Functions of the lunules of *Astriclypeus* as observed in the righting movement (Echinoidea). *Annotnes zool. jap.* **20,** 79-84.

364. Ikegami, S., Tamura, S. and Kanatani, H., 1967. Starfish gonad: Action and chemical identification of spawning inhibitor. *Science, N.Y.* **158,** 1052-1053.

365. Imlay, M. R. and Chaet, A. B., 1965. Microscopic observations of gamete shedding substance in starfish radial nerves. *Fed. Proc.* **24,** 129.

366. Iriye, T. T. and Dille, J. M., 1940. Responses of the isolated longitudinal muscle of *Stichopus californicus* to drugs. *Pharmac. Arch.* **11,** 93-96.

367. Irving, L., 1924. Ciliary tracts propelling haemolymph in *Asterias. J. exp. Zool.* **41,** 115-124.

368. Iwata, K. S. and Fukase, H., 1964a. Artificial spawning in sea urchins by acetylcholine. *Biol. J. Okayama Univ.* **10,** 51-56.

369. Iwata, K. S. and Fukase, H., 1964b. Comparison of discharge of gametes by three artificial means in sea urchins. *Biol. J. Okayama Univ.* **10,** 57-64.

370. JACOBSON, F. W. and MILLOTT, N., 1953. Phenolase and melanogenesis in the coelomic fluid of the echinoid *Diadema antillarum*. *Proc. R. Soc.* B. **141,** 231-247.

371. JENNINGS, H. S., 1907. Behaviour of the starfish *Asterias forreri*. *Univ. Calif. Publs. Zool.* **4,** 53-185.

372. JENSEN, M., 1966. The response of two sea urchins to the sea star *Marthasterias glacialis* and other stimuli. *Ophelia* **3,** 209-219.

373. JEUNIAUX, C., BRICTEUX-GREGOIRE, S. and FLORKIN, M., 1962a. Regulation osmotique intracellulaire chez *Asterias rubens* role du glycocolle et de la taurine. *Cah. Biol. mar.* **3,** 107-113.

374. JEUNIAUX, C., BRICTEUX-GREGOIRE, S. and FLORKIN, M., 1962b. Role osmoregulateur intracellulaire du glycocolle et de la taurine chez l'etoile de mar. *Archs. int. Physiol.* **70,** 155-156.

375. JOHANSEN, K. and VADAS, R. L., 1967. Oxygen uptake and responses to respiratory stress in sea urchins. *Biol. Bull. mar. biol. Lab. Woods Hole* **132,** 16-22.

376. JOHNSON, P. T. and BEESON, R. J., 1966. *In vitro* studies on *Patiria miniata* (Brandt) coelomocytes with remarks on revolving cysts. *Life Sci.* **5,** 1641-1666.

377. JUST, G., 1927. Untersuchungen uber ortsbewegungsreaktionen. I. Das wesen der Phototaktischen reaktionen von *Asterias rubens*. *Z. vergl. Physiol.* **5,** 247-282.

378. KALMUS, H., 1929. Versuche uber die Bewegungen der Seesterne besonders von *Asterina gibbosa*. *Z. vergl. Physiol.* **9,** 703-733.

379. KANATANI, H., 1964. Spawning of starfish: Action of gamete shedding substance obtained from radial nerves *Science, N.Y.* **146,** 1177-1179.

380. KANATANI, H. and NOUMURA, T., 1962. On the nature of active principles responsible for gamete shedding in the radial nerves of starfish. *J. Fac. Sci. Tokyo Univ.*, Ser. 4, **9,** 403-416.

381. KANATANI, H. and NOUMURA, T., 1964. Separation of gamete shedding substance in starfish radial nerves by disc electrophoresis. *Zool. Mag. Tokyo* **73,** 65-69.

382. KANATANI, H. and OHGURI, M., 1966. Mechanism of starfish spawning. I. Distribution of active substance responsible for maturation of oocytes and shedding of gametes. *Biol. Bull. mar. biol. Lab. Woods Hole* **131,** 104-114.

383. KANATANI, H. and SHIRAI, H., 1967. *In vitro* production of meiosis inducing substances by nerve extract in ovary of starfish. *Nature, Lond.* **216,** 284-286.

384. KARANDEEVA, O. G., 1965. Use of an acute experiment to study osmoregulation in aquatic invertebrates. *Ref. Zhur. Biol.* 1965, No. 3D 221.

385. KARNOVSKY, M. L. and BRUMM, A. F., 1955. Studies on naturally occurring α-glycerol ethers. *J. Biol. chem.* **216,** 689-701.

386. KARRER, P., JUCKER, E. and BRAUDE, E. A., 1950. *Carotenoids*. Elsevier, New York.

387. KAWAGUTI, S., 1964. Electron microscopy of the intestinal wall of the sea cucumber with special attention to its muscle and nerve plexus. *Biol. J. Okayama Univ.* **10,** 39-50.

388. KAWAGUTI, S., 1965. Electron microscopy on the radial nerve of the starfish. *Biol. J. Okayama Univ.* **11,** 41-52.

389. KAWAGUTI, S. and KAMISHIMA, Y., 1964. Electron microscope study of the integument of the echinoid *Diadema setosum*. *Annotnes zool. jap.* **37,** 147-152.

390. KAWAMOTO, N., 1926. The anatomy of *Caudina chilensis* with especial reference to the perivisceral coelom, the blood, and the water vascular system in their relation to the blood circulation. *Scient. Rep. Tohoku. Imp. Univ.*, Ser. 4B, **2,** 239-264.

391. KAWAMOTO, N., 1928. Oxygen capacity of the blood of certain invertebrates which contain haemoglobin. *Scient. Rep. Tohoku Imp. Univ.*, Ser 4B, **3,** 561-575.

392. KEČKEŠ, S., 1966. Lunar periodicity in sea cucumbers. *Z. Naturf.* **21,** 1100-1101.

393. KEČKEŠ, S., OZRETIĆ, B. and LUCU, Č., 1966. About a possible mechanism involved in the shedding of sea urchins. *Experientia* **22,** 146-147.

394. KENK, R., 1944. Ecological observations on two Puerto Rican echinoderms, *Mellita lata* and *Astropecten marginatus*. *Biol. Bull. mar. biol. Lab. Woods Hole* **87,** 177-187.

395. KENNEDY, D., 1960. Neural photoreception in a lamellibranch mollusc. *J. gen. Physiol.* **44,** 277-299.

396. KENNEDY, D., 1963. Physiology of photoreceptor neurons in the abdominal nerve cord of the crayfish. *J. gen. Physiol.* **46,** 551-572.

397. KENNEDY, G. Y. and VEVERS, H. G., 1953. Protoporphyrin in the integument of *Asterias rubens*. *Nature, Lond.* **171,** 81.

398. KENNEDY, G. Y. and VEVERS, H. G., 1954. The occurrence of porphyrins in certain marine invertebrates. *J. mar. biol. Ass. U.K.* **33,** 663-676.

399. KERKUT, G. A., 1953. The forces exerted by the tube feet during locomotion. *J. exp. Biol.* **30,** 575-583.

400. KERKUT, G. A., 1954. The mechanisms of co-ordination of the starfish tube feet. *Behaviour* **6,** 206-232.

400b. KERKUT, G. A., 1961. *Implications of Evolution*. Pergamon Press, London.

401. KHAILOV, K. M., 1962. The use of phenol extraction in the investigation of the organic complex of sea water. *Okeanologiya* **2,** 835-844.

402. KINDRED, J. E., 1921. Phagocytosis and clotting in the perivisceral fluid of *Arbacia*. *Biol. Bull. mar. biol. Lab. Woods Hole* **41,** 144-152.

403. KINDRED, J. E., 1924. The cellular elements in the perivisceral fluid of echinoderms. *Biol. Bull. mar. biol. Lab. Woods Hole* **46,** 228-251.

404. KINOSHITA, H., 1941. Conduction of impulse in superficial nervous system of sea urchin. *Jap. J. Zool.* **9,** 221-232.

405. KITTREDGE, J. S., SIMONSEN, D. G., ROBERTS, E. and JELINCK, B., 1962. Free amino acids of marine invertebrates, in *Amino Acid Pools*, Ed. J. T. Holden, Elsevier, Amsterdam, pp. 176-186.

406. KJERSKOG-AGERSBORG, H. P., 1918. Bilateral tendencies and habits in the 20-rayed starfish *Pycnopodia helianthoides*. *Biol. Bull. mar. biol. Lab. Woods Hole* **35,** 232-254.

407. KJERSKOG-AGERSBORG, H. P., 1922. The relation of the madreporite to the physiological anterior end in the 20-rayed starfish *Pycnopodia helianthoides*. *Biol. Bull. mar. biol. Lab. Woods Hole* **42,** 202-216.

408. KLEINHOLTZ, L. K., 1938. Color changes in echinoderms. *Pubbl. Staz. zool. Napoli* **17,** 53-57.

409. KOBAYASHI, S., 1932. The spectral properties of haemoglobin in the holothurian *Caudina chilensis* and *Molpadia roretzii*. *Scient. Rep. Tohoku Imp. Univ.*, Ser. 4 **7**, 211-227.

410. KOHN, A. J., 1961. Chemoreception in gastropod molluscs. *Am. Zool.* **1**, 291-308.

411. KOIZUMI, T., 1932. Studies on the exchange and the equilibrium of water and electrolytes in a holothurian *Caudina chilensis* (Muller). I. Permeability of the animal surface to water and ions in sea water together with osmotic and ionic equilibrium between the body fluid of the animal and its surrounding sea water, involving some corrections to our previous paper. *Scient. Rep. Tohoku Imp. Univ.*, Ser. 4 **7**, 259-311.

412. KOIZUMI, T., 1935a. Studies on the exchange and the equilibrium of water and electrolytes in a holothurian *Caudina chilensis* (Muller). II. Velocity of permeation of chloride and sulphate through the isolated body wall of *Caudina*. *Scient. Rep. Tohoku Imp. Univ.*, Ser. 4 **10**, 33-39.

413. KOIZUMI, T., 1935b. Studies on the exchange and the equilibrium of water and electrolytes in a holothurian *Caudina chilensis* (Muller). III. On the velocity of permeation of potassium, sodium, calcium, and magnesium through isolated body wall. *Scient. Rep. Tohoku Imp. Univ.*, Ser. 4 **10**, 269-275.

414. KOIZUMI, T., 1935c. Studies on the exchange and the equilibrium of water and electrolytes in a holothurian *Caudina chilensis* (Muller). IV. On the inorganic composition of the corpuscles of the body fluid. *Scient. Rep. Tohoku Imp. Univ.*, Ser. 4 **10**, 277-281.

415. KOIZUMI, T., 1935d. Studies on the exchange and the equilibrium of water and electrolytes in a holothurian *Caudina chilensis* (Muller). V. Inorganic composition of the longitudinal muscles and of the body wall without the longitudinal muscles. *Scient. Rep. Tohoku Imp. Univ.*, Ser. 4 **10**, 281-286.

416. KOLLER, G., 1930. Versuche an marinen Wirbellosen uber die Aufnahme geloster Nahrstoffe. *Z. vergl. Physiol.* **11**, 437-447.

417. KOLLER, G. and MEYER, H., 1933. Versuche uber die Atmung der Echinodermen. *Biol. Zbl.* **53**, 655-661.

418. KONYSHEV, V. A. and MURASHOVA, A. I., 1965. Primary immunological reactivity of echinoderms and ascidians. *Zhur. Obshch. Biol.* **25**, 451-457.

419. KORRINGA, P., 1951. The shell of *Ostrea edulis* as a habitat. *Archs. neerl. Zool.* **10**, 32-152.

420. KOSSEL, A. and EDELBACHER, S., 1915. Beitrage zur Kenntnis der Echinodermen. *Hoppe-Seyler's Z. physiol. Chem.* **94**, 264-283.

421. KOWALEVSKY, A., 1889. Ein beitrage zur Kenntnis der Exkretionsorgane. *Biol. Zbl.* **9**, 65-76.

422. KOWALSKI, R., 1955. Untersuchungen zur biologie des Seesternes *Asterias rubens* L. in Brachwasser. *Kieler Meeresforsch.* **11**, 201-212.

423. KRISTENSEN, I., 1964. The effect of raising *Diadema* at different levels of light intensity on pigmentation and preference for darkness. *Caribb. J. Sci.* **4**, 441-443.

424. KRUGER, F., 1932. Versuche uber die Wasserbewegung durch die Madreporenplatte von *Echinus*. *Z. vergl. Physiol.* **18**, 157-173.

425. KRUKENBURG, C. F. W., 1882. Das Gorgonidenroth und Britische Bemerkungen zu Merejowski angebliche Endeckungen des Zooerythrines

bei wirbellosen Tieren. *Vergleichend-physiologische Studien*, Ser. 2, 92-93. Carl Winter, Heidelburg.

426. KUBO, K., 1959. Studies on the systematic serology of sea stars. IV. *Annotnes. zool. jap.* **32,** 214-219.

427. KUHL, W., 1965. Das bewegungsverhalten der Coelomzellen von *Psammechinus miliaris* bei der Wundheibing (Echinodermata). *Helg. wiss. Meeres.* **12,** 424-443.

428. KUHN, R. and WALLENFELS, K., 1940. The spine pigment of *Arbacia. Ber. dtsch. chem. Ges.* **74B,** 1594-1598.

429. KUHN, R., WALLENFELS, K., WEYGAND, F., MOLL, T. and HEPDING, L., 1939. The specificity of vitamin K. *Naturwissenschaften* **27,** 518-519.

430. KURODA, C. and HARADA, M., 1955. The pigments from the sea urchins and the syntheses of the related compounds. VIII. *Proc. imp. Acad. Japan* **31,** 305-308.

431. KURODA, C. and INAKURA, H., 1942. The pigments from the sea urchins and the syntheses of the related compounds. II. *Proc. imp. Acad. Japan* **18,** 74.

432. KURODA, C. and KOYASU, K., 1944. Studies on the derivatives of naphthoquinone. *Proc. imp. Acad. Japan* **20,** 23-25.

433. KURODA, C. and OKAJIMA, M., 1950. Studies on the derivatives of naphthoquinone. *Proc. imp. Acad. Japan* **26,** 33-36.

434. KURODA, C. and OKAJIMA, M., 1951. Studies on the derivatives of naphthoquinone. *Proc. imp. Acad. Japan* **27,** 343-345.

435. KURODA, C. and OKAJIMA, M., 1953. Studies on the derivatives of naphthoquinone. *Proc. imp. Acad. Japan* **29,** 27-29.

436. KURODA, C. and OKAJIMA, M., 1954. Studies on the derivatives of naphthoquinone. *Proc. imp. Acad. Japan* **30,** 982-986.

437. KURODA, C. and OKAJIMA, M., 1958. Studies on the derivatives of naphthoquinone. *Proc. imp. Acad. Japan* **34,** 616-618.

438. KURODA, C. and OKAJIMA, M., 1960. Studies on the derivatives of naphthoquinone. *Proc. imp. Acad. Japan* **36,** 424-427.

439. KURODA, C. and OKAJIMA, M., 1962. Studies on the derivatives of naphthoquinone. *Proc. imp. Acad. Japan* **38,** 353-355.

440. KURODA, C. and OSHIMA, H., 1940. The pigments from the sea urchins and the syntheses of related compounds. *Proc. imp. Acad. Japan* **16,** 214-217.

441. KUTSCHER, F. and ACKERMANN, D., 1926. Vergleichendphysiologie Untersuchungen von Extrakten verscheidener tierklassen auf tierische alkaloide eine Zusammerfassung. *Z. Biol.* **84,** 181-192.

442. KUWADA, S. and BAN, S., 1949. Studies on steroids. XXVIII. Patiriasterol from starfish in Japan. *J. Pharmac. soc. Japan* **69,** 10-13.

443. LANDENBERGER, D. E., 1966. Learning in the Pacific starfish *Pisaster giganteus. Anim. Behav.* **14,** 414-418.

444. LANGE, R., 1963. The osmotic function of amino acids and taurine in the mussel *Mytilus edulis. Comp. Biochem. Physiol.* **10,** 173-179.

445. LANGE, R., 1964. The osmotic adjustment in the echinoderm *Strongylocentrotus droebachiensis. Comp. Biochem. Physiol.* **13,** 205-216.

446. LANGE, W., 1876. Beitrage zur Anatomie und Histologie der Asterien und Ophiuren. *Morph. Jb.* **2,** 241-286.

447. LANGELOT, H.-P., 1937. Uber die Bewegungen von *Antedon rosaceous* und ihre nervose Regulierung. *Zool. Jb.* **59,** 235-279.

448. LANKESTER, E. R., 1900. *A Treatise on Zoology. III. The Echinodermata.* Black, London, 344 pp.
449. LASKER, R. and BOOLOOTIAN, R. A., 1960. Digestion of the alga *Macrocystis pyrifera* by the sea urchin *Strongylocentrotus purpuratus. Nature, Lond.* **188,** 1130.
450. LASKER, R. and GIESE, A. C., 1954. Nutrition of the sea urchin *Strongylocentrotus purpuratus. Biol. Bull. mar. biol. Lab. Woods Hole* **106,** 328-340.
451. LAVOIE, M. E., 1956. How sea stars open bivalves. *Biol. Bull. mar. biol. Lab. Woods Hole* **111,** 114-122.
452. LAWRENCE, D. C., LAWRENCE, A. L., GREER, M. L. and MAILMAN, D., 1967. Intestinal absorption in the sea cucumber *Stichopus parvimensis. Comp. Biochem Physiol.* **20,** 619-627.
453. LAWRENCE, J. M., LAWRENCE, A. L. and GIESE, A. C., 1966. Role of the gut as a nutrient-storage organ in the purple sea urchin *Strongylocentrotus purpuratus. Physiol. Zool.* **39,** 281-290.
454. LAWRENCE, J. M., LAWRENCE, A. L. and HOLLAND, N. D., 1965. Annual cycle in the size of the gut of the purple sea urchin *Strongylocentrotus purpuratus. Nature, Lond.* **205,** 1238-1239.
455. LEDERER, E., 1935. Echinenone et pentaxanthine: deux nouveaux carotenoides trouves dans l'oursin (*Echinus esculentus*). *C. r. hebd. Séanc. Acad. Sci. Paris* **201,** 300-302.
456. LEDERER, E., 1938. Recherches sur les carotenoides des Invertebrès. *Bull. Soc. Chim. biol.* **20,** 567-610.
457. LEDERER, E., 1940. Les pigments des Invertebrès. *Biol. Rev.* **15,** 273-306.
458. LEDERER, E., 1952. Sur les pigments naphthoquinones des epines et du test des Oursins *Paracentrotus lividus* et *Arbacia pustulosa. Biochim. biophys. Acta* **9,** 92-101.
459. LEDERER, E. and GLASER, R., 1938. Sur l'echinochrome et le spinochrome. *C. r. hebd. Séanc. Acad. Sci. Paris* **207,** 454.
460. LEDERER, E. and MOORE, T., 1936. Echinenone as a Pro-vitamin A. *Nature, Lond.* **137,** 996.
461. LEE, C. F., 1951. Technological studies of the starfish. *Fish. Leafl. Wash.* **391,** 1-47.
462. LEVIN, A. and WYMAN, J., 1927. The viscous elastic properties of muscle. *Proc. R. Soc.* B. **101,** 218-243.
463. LEWIS, J. B., 1958. The biology of the tropical sea urchin *Tripneustes esculentus* Leske in the Barbados, B.W.I. *Can. J. Zool.* **36,** 607-621.
464. LEWIS, J. B., 1963. The food of some deep water echinoids from Barbados. *Bull. mar. Sci. Gulf Caribb.* **13,** 360-363.
465. LEWIS, J. B., 1964. Feeding and digestion in the tropical sea urchin *Diadema antillarum. Can. J. Zool.* **42,** 549-557.
466. LEWIS, J. B., 1966. Growth and breeding in the tropical echinoid *Diadema antillarum. Bull. mar. Sci. Gulf Caribb.* **16,** 151-158.
467. LEWIS, J. B., 1967. Nitrogenous excretion in the tropical sea urchin *Diadema antillarum* Phillipi. *Biol. Bull. mar. biol. Lab. Woods Hole* **132,** 34-37.
468. LIEBMANN, E., 1947. The trephocytes and their function. *Experientia* **3,** 442-451.
469. LIEBMANN, E., 1950. The leucocytes of *Arbacia punctulata. Biol. Bull. mar. biol. Lab. Woods Hole* **98,** 46-59.

470. LINDAHL, P. E. and RUNNSTROM, J., 1929. Variation und Okologie von *Psammechinus miliaris* (Gmelin). *Acta zool. Stockh.* **10,** 401-484.
471. LISON, L., 1930. Recherches histophysiologiques sur les amibocytes des Echinodermes. *Archs. Biol. Paris* **40,** 175-203.
472. LOEB, J., 1900. *Comparative Physiology of the Brain and Comparative Psychology.* Putnam, New York.
473. LONNBERG, E., 1931. Untersuchungen uber das Vorkommen carotinoiden Stoffe bei marinen Evertebraten. *Ark. Zool.* **22A**, No. 14, 1-49.
474. LONNBERG, E., 1933. Zur Kenntniss der Carotinoide bei marinen Evertebraten. *Ark. Zool.* **25A,** No. 1, 1-17.
475. LONNBERG, E., 1934. Weitere Beitrage zur Kenntnis der Carotinoide der marinen Evertebraten. *Ark. Zool.* **26A,** No. 7, 1-36.
476. LONNBERG, E. and HELLSTROM, H., 1932. Zur kenntnis der Carotinoide bei marinen Evertebraten. *Ark. Zool.* **23A,** No. 15, 1-74.
477. LOOSANOFF, V., 1945. Effect of sea water of reduced salinities upon *Asterias forbesi* of Long Island Sound. *Trans. Conn. Acad. Arts. Sci.* **36,** 813-833.
478. LUDWIG, H., 1890. Uber die funktion der madreporenplatte und die steincanals der Echinodermen. *Zool. Anz.* **13,** 377-379.
479. LUTZ, B. R., 1930. The effect of low oxygen tension on the pulsation of the isolated holothurian cloaca. *Biol. Bull. mar. biol. Lab. Woods Hole* **58,** 74-84.
480. LYONS, R. B., BISHOP, W. R. and BACON, R. L., 1964. Fine structure of the scalloped flagellated cup cell of the sea urchin *Strongylocentrotus purpuratus. J. Cell. Biol.* **23,** 55A.
481. MAGAZANIK, L. G., 1965. Qualitative characteristics of the choline receptors of muscles in some deuterostomatic animals (holothurians, ascidians, frogs). *Zhur. Evolyuts Biokhim. Fiziol.* **1,** 220-226.
482. MAGAZANIK, L. G. and FRUENTOV, N. K., 1963. The evolution of cholinoceptive sites of locomotor muscles. *Biochem. Pharmac.* **12,** 52.
483. MALOEUF, N. S. R., 1938. Studies on the respiration and osmoregulation of animals. *Z. vergl. Physiol.* **125,** 1-28.
484. MANGOLD, E., 1908a. Studien zur physiologie des Nervensystems der Echinodermen. I. Die Fusschen der Seesterne und die Co-ordination ihrer Bewegungen. *Pflugers Arch. ges. Physiol.* **122,** 315-360.
485. MANGOLD, E., 1908b. Studien zur physiologie des Nervensystems der Echinodermen. II. Uber das Nervensystems der Seesterne und uber den tonus. *Pflugers Arch. ges. Physiol.* **123,** 1-39.
486. MANGOLD, E., 1909. Sinnesphysiologisches Untersuchungen an Echinodermen. Ihre Reactionen auf Licht und Schatten und die negative Geotaxis bei *Asterina. Z. allg. Physiol.* **9,** 112-146.
487. MANGOLD, E., 1921. Der umdrehreflex bei seesternen und Schlangensternen. *Pflugers Arch. ges. Physiol.* **189,** 73-108.
488. MANUNTA, C., 1943. Richerche su la pigmentazione di Oloturie, Spugne ed altri animali marini delle Coste istrione. *Thalassia* **5,** No. 5, 1-28.
489. MANUNTA, C., 1944. Sui pigmenti carotenoidi di invertibrati marini delle Coste di Rovigno d'Istria. *Thalassia* **6,** No. 3, 1-28.
490. MANUNTA, C., 1947. Contenuto in lipidi negli organi viscerali di tre speci di Olothurie delle Coste di Rovigno d'Istria. *Archo. zool. ital.* **32,** 25-31.
491. MANWELL, C., 1959. Oxygen equilibrium of *Cucumaria miniata* haemoglobin and the absence of the Bohr effect. *J. cell. comp. Physiol.* **53,** 75-83.

492. MANWELL, C., 1964. The blood proteins of cyclostomes: A study in phylogenetic and ontogenetic biochemistry, in *The Biology of Myxine*, Ed. A. Brodal and R. Fange. Universitetsforlaget, 372-456.

493. MANWELL, C., 1966. Sea cucumber sibling species: polypeptide chain types and oxygen equilibrium of haemoglobin. *Science, N.Y.* **152,** 1393-1396.

494. MANWELL, C. and BAKER, C. M. A., 1963. A sibling species of sea cucumber discovered by starch gel electrophoresis. *Comp. Biochem. Physiol.* **10,** 39-53.

495. MARGOLIN, A. B., 1964. The mantle response of *Diadora aspersa. Anim. Behav.* **12,** 187-194.

496. MARKEL, K. and TITSHACH, H., 1965. Das Festholtevermögen von Seeigeln und die Reissfestigheit ihrer Ambulacralfüsschen. *Naturwissenschaften* **52,** 268.

497. MARSHALL, A. M., 1884. On the nervous system of *Antedon rosaceous. Q. Jl Microsc. Sci.* **24,** 507-548.

498. MARUYAMA, K. and MATSUMIYA, H., 1957. The contractile protein from tube feet of a starfish. *J. Biochem. Tokyo* **44,** 537-542.

499. MATSUMOTO, T. and TAMOURA, T., 1956. The unsaponifiable steroid substances of Echinoderms. 3. Brittle stars. *Nippon Kaguku Zashi.* **77,** 376-378.

500. MATSUMOTO, T. and TOYAMA, Y., 1943. Sterols and other unsaponifiable substances in fats of shell fishes Crustacea and Echinodermata. IV. Separation of a new sterol from the fat of the starfish *Asterina pectinifera. J. Chem. Soc. Japan* **64,** 1069-1071.

500b. MATSUMOTO, T., WAINAI, T. and HIRAI, C., 1956. Unsaponifiable matter of echinoderms. Δ^7 stigmasterol and stellasterol in the fat of the starfish. *Nippon Kagaku Zashi* **77,** 531-533.

501. MAUZEY, K. P., 1966. Feeding behaviour and reproductive cycles in *Pisaster ochraceous. Biol. Bull. mar. biol. Lab. Woods Hole* **131,** 127-144.

502. MAY, R. M., 1925. Les reactions sensorielles d'une ophiure. *Biol. Bull. Fr. et Belg.* **59,** 372-402.

503. MEAD, A. D., 1901. The natural history of the starfish. *Bull. U.S. Fish. Commn.* **19,** 203-224.

504. MECKLENBURG, T. A. and CHAET, A. B., 1964. Calcium and the shedding substances of *Patiria miniata. Am. Zool.* **4,** 414-415.

505. MENDES, E. G., ABBUD, L. and UMIJI, S., 1963. Cholinergic action of homogenates of sea urchin pedicellariae. *Science, N.Y.* **139,** 408-409.

506. MESSINA, L., 1958. Prime osservazione sulla natura chimica del liquido perivisciale di *Arbacia lixula. Pubbl. Staz. zool. Napoli* **30,** 127-131.

507. METTRICK, D. F. and TELFORD, J. M., 1963. Histamine in non-vertebrate animals. *J. Pharm. Pharmac.* **15,** 694-697.

508. METTRICK, D. F. and TELFORD, J. M., 1965. The histamine content and histidine decarboxylase activity of some marine and terrestrial animals from the West Indies. *Comp. Biochem. Physiol.* **16,** 547-559.

509. MEYER, H., 1935. Die Atmung von *Asterias rubens* und ihre Abhandigskeit von verscheidenen ausfaktoren. *Zool. Jb.* **55,** 349-398.

510. MEYER, J. A., 1914. Beitrage zur Kenntnis der chemischen Zusammensetzung wirbelloser Tiere. *Wiss. Meeresuntersuch.* **16,** 231-283.

511. MILLIGAN, H. N., 1916. Rate of locomotion in sun-stars. *Zoologist*, **20,** 319.

512. MILLOTT, N., 1950a. The sensitivity to light, reactions to shading, pigmenta-

tion and color change of the sea urchin *Diadema antillarum*. *Biol. Bull. mar. biol. Lab. Woods Hole* **99,** 329.

513. MILLOTT, N., 1950b. Integumentary pigmentation and the coelomic fluid of *Thyone briareus*. *Biol. Bull. mar. biol. Lab. Woods Hole* **99,** 343-344.
514. MILLOTT, N., 1952a. Colour change in the echinoid *Diadema antillarum*. *Nature, Lond.* **170,** 325-326.
515. MILLOTT, N., 1952b. The occurrence of melanin and phenolases in *Holothuria forskali*. *Experientia* **8,** 301.
516. MILLOTT, N., 1953a. Observations on the skin pigment and amoebocytes and the occurrence of phenolases in the coelomic fluid of *Holothuria forskali*. *J. mar. biol. Ass. U.K.* **31,** 529-540.
517. MILLOTT, N., 1953b. Light emission and light perception in a species of *Diadema*. *Nature, Lond.* **171,** 973-974.
518. MILLOTT, N., 1953c. Some preliminary observations on the young forms of the echinoid *Diadema antillarum*. *Bull. mar. Sci. Gulf. Caribb.* **2,** 497-510.
519. MILLOTT, N., 1954a. A note on the shedding of pigment from the body wall of *Holothuria grisea*. *Bull. mar. Sci. Gulf. Caribb.* **3,** 305-306.
520. MILLOTT, N., 1954b. Sensitivity to light and the reactions to changes in light intensity of the echinoid *Diadema antillarum* Phillipi. *Phil. Trans. R. Soc.* B. 238, 187-220.
521. MILLOTT, N., 1956. The covering reaction of sea urchins. I. A preliminary account of covering in the tropical echinoid *Lytechinus variegatus* and its relation to light. *J. exp. Biol.* **33,** 508-523.
522. MILLOTT, N., 1957. Naphthoquinone pigment in the tropical sea urchin *Diadema antillarum* (Phillipi). *Proc. zool. Soc. Lond.* **129,** 263-272.
523. MILLOTT, N., 1961. The photosensitivity of sea urchins, in *Comparative Biochemistry of Photoreactive Systems*, Ed. M. B. Allen, Academic Press, New York.
524. MILLOTT, N., 1964. The pigmentary system of *Diadema antillarum*. *Nature, Lond.* **203,** 206-207.
525. MILLOTT, N., 1966. A possible function for the axial organ of Echinoids. *Nature, Lond.* **209,** 594-596.
526. MILLOTT, N., 1967a. Dermal sensitivity and the "Hen and Egg" problem. *Nature, Lond.* **215,** 768-769.
527. MILLOTT, N., 1967b. The axial organ of echinoids: Re-interpretation of its structure and function, in *Echinoderm Biology*, Ed. N. Millott. Symp. zool. Soc. London., No. 20.
528. MILLOTT, N. and FARMANFARMAIAN, A., 1967. Regeneration of the axial organ of *Arbacia punctulata* and its implications. *Nature, Lond.* **216,** 1136-1138.
529. MILLOTT, N. and JACOBSON, F. W., 1951. Phenolases in the echinoid *Didema antillarum*. *Nature, Lond.* **168,** 878.
530. MILLOTT, N. and MANLY, B. M., 1961. The iridophores of the echinoid *Diadema antillarum*. *Q. Jl Microsc. Sci.* **102,** 181-194.
531. MILLOTT, N. and OKAMURA, H., 1968a. Pigmentation in the radial nerve of *Diadema antillarum*. *Nature, Lond.* **217,** 92-93.
532. MILLOTT, N. and OKAMURA, H., 1968b. The electrical activity of the radial nerve in *Diadema antillarum* Phillipi and certain other echinoids. *J. exp. Biol.* **48,** 279-287.

533. Millott, N. and Takahashi, K., 1963. The shadow reaction of *Diadema antillarum*. IV. Spine movements and their implications. *Phil. Trans. R. Soc.*, B. **246,** 437-469.
534. Millott, N. and Vevers, H. G., 1955. Carotenoid pigments in the optic cushion of *Marthasterias glacialis*. *J. mar. biol. Ass. U.K.* **34,** 279-287.
535. Millott, N. and Vevers, H. G., 1964. Axial organ and fluid circulation in echinoids. *Nature, Lond.* **204,** 1216-1217.
536. Millott, N. and Vevers, H. G., 1968. The morphology and histochemistry of the echinoid axial organ. *Phil. Trans. R. Soc.*, B. **253,** 201-230.
537. Millott, N. and Yoshida, M., 1957. The spectral sensitivity of the echinoid *Diadema antillarum*. *J. exp. Biol.* **34,** 394-401.
538. Millott, N. and Yoshida, M., 1959. The photosensitivity of the sea urchin *Diadema antillarum*: Responses to increase in light intensity. *Proc. zool. Soc. Lond.* **133,** 67-71.
539. Millott, N. and Yoshida, M., 1960a. The shadow reaction of *Diadema antillarum*. I. The spine response and its relation to the stimulus. *J. exp. Biol.* **37,** 363-375.
540. Millott, N. and Yoshida, M., 1960b. The shadow reaction of *Diadema antillarum*. II. Inhibition by light. *J. exp. Biol.* **37,** 376-389.
541. Monks, S. P., 1904. Variability and autotomy of *Phataria*. *Proc. Acad. nat. Sci. Philad.* **55,** 596-600.
542. Montouri, A., 1913. Les processus oxydatifs chez les animaux marins en rapport avec la loi de superficie. *Archs. ital. Biol.* **59,** 213-234.
543. Moore, A. R., 1910a. On the righting movements of the starfish. *Biol. Bull. mar. biol. Lab. Woods Hole* **19,** 235-239.
544. Moore, A. R., 1910b. On the nervous mechanism of the righting movements of the starfish. *Am. J. Physiol.* **27,** 207-211.
545. Moore, A. R., 1918. Reversal of reaction by means of Strychnine in planarians and starfish. *J. gen. Physiol.* **1,** 97-100.
546. Moore, A. R., 1920. Stereotropism as a function of neuromuscular organisation. *J. gen. Physiol.* **2,** 319-324.
547. Moore, A. R., 1921. Stereotropic orientation of the tube feet of starfish (*Asterias*) and its inhibition by light. *J. gen. Physiol.* **4,** 163-169.
548. Moore, A. R., 1924. The nervous mechanism of co-ordination in the crinoid *Antedon rosaceous*. *J. gen. Physiol.* **6,** 281-288.
549. Moore, H. B., Jutare, T., Jones, J. A. and McPherson, B. F., 1963. A contribution to the biology of *Tripneustes esculentus*. *Bull. mar. Sci. Gulf Caribb.* **13,** 267-281.
550. Moore, H. B. and Lopez, N. N., 1966. The ecology and productivity of *Moira atropos*. *Bull. mar. Sci. Gulf Caribb.* **16,** 648-667.
551. Moore, H. B. and McPherson, B. F., 1965. A contribution to the study of the productivity of the urchins *Tripneustes esculentus* and *Lytechinus variegatus*. *Bull. mar. Sci. Gulf Caribb.* **15,** 855-871.
552. Moreland, B., Watts, D. C. and Virden, R., 1967. Phosphagen kinases and evolution in the Echinodermata. *Nature, Lond.* **214,** 458-462.
553. Mori, S. and Matsutani, K., 1952. Studies on the daily rhythmic activity of the starfish *Astropecten polyacanthus* Muller et Troschel, and the accompanied physiological rhythms. *Publ. Seto mar. biol. Lab.* **2,** 213-225.

554. Mortensen, T., 1927. *Handbook of the Echinoderms of the British Isles.* Oxford Univ. Press, London.
555. Mortensen, T., 1937. Contributions to the study of the development and larval forms of echinoderms. III. *K. danske Vidensk. Selsk. Natur. Math. Afd.*, Ser 9, **7,** 1-65.
556. Mortensen, T., 1938. Contributions to the study of the development and larval forms of echinoderms. IV. *K. danske Vidensk. Selsk. Natur. Math. Afd.*, Ser. 9. **7,** 1-59.
557. Mosley, H. N., 1877. On the colouring matters of various animals and especially of deep sea forms dredged by H.M.S. *Challenger. Q. Jl Microsc. Sci.* **17,** 1-23.
557b. Motohiro, T., 1960. Studies on mucoprotein in marine products. *Bull. Jap. Soc. scient. Fish.* **26,** 1171-1182.
558. Muntz, L., Ebling, F. J. and Kitching, J. A., 1965. The ecology of Lough Ine. III. Predatory activity of large crabs. *J. Anim. Ecol.* **34,** 315-329.
559. Musajo, L. and Minchilli, M., 1942. Su di un secondo pigmento degli aculei di paracentrotus. *Boll. Sci. Fac. Chim. ind. Univ. Bologna* **31,** 113-116.
560. Myers, R. G., 1920. A chemical study of the blood of several invertebrate animals. *J. biol. Chem.* **41,** 119-135.
561. McBride, E. W., 1896. The development of *Asterina gibbosa. Q. Jl Microsc. Sci.* **38,** 339-412.
562. MacCurdy, H., 1912. Observations on the reactions of *Asterias forbesi* to light. *Science, N.Y.* **35,** 192.
563. MacCurdy, H., 1913. Some effects of sunlight on the starfish. *Science, N.Y.* **38,** 98-100.
564. MacGinitie, G. E., 1939. Some effects of freshwater on the fauna of a marine harbour. *Am. Midl. Nat.* **21,** 681-686.
565. MacGinitie, G. E. and MacGinitie, N., 1949. *Natural History of Marine Animals.* McGraw-Hill, New York, 473 pp.
566. MacMunn, C. A., 1883. Studies in animal chromatology. *Proc. Bgham philos. Soc.* **3,** 351-407.
567. MacMunn, C. A., 1885. On the chromatology of the blood of some invertebrates. *Q. Jl Microsc. Sci.* **25,** 469-490.
568. MacMunn, C. A., 1887. On the presence of haematoporphyrin in the integument of certain vertebrates. *J. Physiol. Lond.* **7,** 240-252.
569. MacMunn, C. A., 1889. Contributions to animal chromatology. *Q. Jl Microsc. Sci.* **30,** 51-96.
570. Nagabhushanam, A. K. and Colman, J. S., 1959. Carrion eating by ophiuroids. *Nature, Lond.* **184,** 285.
571. Nakazawa, S., 1960. Dissociation of animal tissues by a toxic substance obtained from starfish. *Naturwissenschaften* **47,** 327-328.
572. Needham, D. M., Needham, J., Baldwin, E. and Yudkin, J., 1932. A comparative study of the phosphagens with some remarks on the origin of vertebrates. *Proc. R. Soc.* B. **110,** 260-294.
573. Nelson, J. R., 1946. Observations on the occurrence and possible uses of starfish. *Bull. Bingham oceanogr. Coll.* **9,** Art. 3, 31-32.
574. Newell, R. C. and Courtney, W. A. M., 1965. Respiratory movements in *Holothuria forskali* Delle chiajei. *J. exp. Biol.* **42,** 45-57.

575. NICHOLS, D., 1959a. Histology of the tube feet and clavulae of *Echinocardium cordatum. Q. Jl Microsc. Sci.* **100,** 73-87.
576. NICHOLS, D., 1959b. Histology and activities of the tube feet of *Echinocyamus pusillus. Q. Jl Microsc. Sci.* **100,** 539-555.
577. NICHOLS, D., 1960. The histology and activities of the tube feet of *Antedon bifida. Q. Jl Microsc. Sci.* **101,** 105-117.
578. NICHOLS, D., 1961. A comparative histological study of the tube feet of two regular echinoids. *Q. Jl Microsc. Sci.* **102,** 157-180.
579. NICHOLS, D., 1962. *Echinoderms.* Hutchinson, London, 200 pp.
580. NICHOLS, D., 1964. Echinoderms: Experimental and Ecological. *Ann. Rev. oceanogr. mar. Biol.* **2,** 393-423.
581. DINICOLA, M., 1954. The carotenoids of the carapace of the echinoderm *Ophidiaster ophidianus. Biochem. J.* **56,** 555-558.
582. DINICOLA, M., 1956. Astaxanthin in asteroid echinoderms: *Asterina panceri. Expl. Cell Res.* **10,** 441-446.
583. DINICOLA, M., 1958. Il carotenoidi dell integumento degli Asteroidi: *Astropecten aurantiacus. Ricerca Sci.* **28,** 2076-2083.
584. DINICOLA, M., 1959. Il carotenoidi negli Asteroidi: *Marthasterias glacialis. Ricerca Sci.* **29,** 78-84.
585. DINICOLA, M. and GOODWIN, T. W., 1954. The distribution of carotenoids in some marine invertebrates. *Pubbl. Staz. zool. Napoli* **25,** 145-160.
586. NIGRELLI, R. F., 1952. The effects of holothurin on fish and mice with Sarcoma 180. *Zoologica, N.Y.* **37,** 89-90.
587. NIGRELLI, R. F., CHANLEY, J. D., KOHN, S. K. and SOBOTKA, H., 1955. The chemical nature of holothurin, a toxic principle from the sea cucumber (Echinodermata, Holothuroidea). *Zoologica, N.Y.* **40,** 47-48.
588. NIGRELLI, R. F. and JAKOWSKA, S., 1960. Effects of holothurin, a steroid saponin from the Bahamian sea cucumber *Actinopyga agassizi,* on various biological systems. *Ann. N.Y. Acad. Sci.* **90,** Art 3, 884-892.
589. NISHIBORI, K., 1953. Studies on marine animal pigments. II. On the pigment components from skins of Asteroidea. *Bull. Jap. Soc. scient. Fish.* **19,** 648.
590. NISHIBORI, K., 1959. Isolation of Echinochrome A from the spines of the sea urchin *Diadema setosum. Nature, Lond.* **184,** 1234.
591. NOLL, F. C., 1881. Mein Seewasser-Zimmeraquarium. *Der Zool. Garten.* **22,** 168-177.
592. NOMURA, S., 1926. The influence of oxygen tension on the rate of oxygen consumption in *Caudina. Scient. Rep. Tohoku Imp. Univ.,* Ser. 4, **2,** 133-138.
593. NOUMURA, T. and KANATANI, H., 1962. Induction of spawning by radial nerve extracts in some starfishes. *J. Fac. Sci. Tokyo Univ.,* Sect. 4, **9,** 397-402.
594. NYHOLM, K.-G., 1951. The development and larval form of *Labidoplax buskii. Zool. Bidr.* **29,** 239-254.
595. O'DONOGHUE, C. H., 1924. On the summer migration of certain starfish in Departure Bay, B.C. *Contrib. Can. Biol.,* N.S. **1,** 457-472.
596. OGAWA, S., 1928. A new apparatus used for the study of respiration in *Caudina chilensis. Scient. Rep. Tohoku Imp. Univ.,* Ser. 4. **3,** 39-49.
597. OHSHIMA, H., 1940. The righting movements of the sea star *Oreaster nodosus. Jap. J. Zool.* **8,** 575-589.
598. OHUYE, T., 1937. On the coelomic corpuscles in the body fluid of some

invertebrates. IV. On the coelomic corpuscles of a holothuroid *Molpadia roretzii* with reference to those of *Caudina chilensis. Scient. Rep. Tohoku Imp. Univ.*, Ser. 4. **11,** 207-222.

599. OHUYE, T., 1938. On corpuscles in the body fluids of some invertebrates. *Scient. Rep. Tohoku Imp. Univ.*, Ser. 4. **13,** 359-380.

600. OKADA, K., 1955. Biological studies on the practical utilities of poisonous marine invertebrates. I. A preliminary note on the toxical substance detected in the trumpet sea urchin *Toxopneustes pileolus. Recent oceanogr. wks. Japan* **2,** 49-54.

601. OKADA, K., HASHIMOTO, T. and MIYAUCHI, Y., 1955. A preliminary report on the poisonous effect of the *Toxopneustes* toxin upon the heart of oyster. *Bull. mar. biol. Sta. Asamushi* **7,** 133-140,

602. OKAZAKI, K., 1956. Skeleton formation in sea urchin larvae. I. Effect of calcium concentration. *Biol. Bull. mar. biol. Lab. Woods Hole* **110,** 320-333.

603. OKAZAKI, K., 1960. Skeleton formation of sea urchin larvae. II. Organic matrix of the spicule. *Embryologia* **5,** 283-320.

604. OLMSTED, J. M., 1917. The comparative physiology of *Synaptula hydriformis. J. exp. Zool.* **24,** 333-379.

605. OLSEN, H., 1942. The development of the brittle star *Ophiopholis aculeata* (O. Fr. Muller) with a short report on the outer hyaline layer. *Univ. Bergen Arb.* **6,** 5-107.

606a. OOMEN, H. A. P. C., 1926. Permeability of the gut in sea cucumbers. *Kon. Acad. Wetens, Amsterdam, Natur. Afd.* **28,** 1000-1007.

606b. OOMEN, H. A. P. C., 1926b. Verdaungsphysiologische Studien an Holothurien. *Pubbl. Staz. zool. Napoli* **7,** 215-297.

607. ORR, P. R., 1955. Heat death. I. Time-temperature relationships in marine animals. *Physiol. Zool.* **28,** 290-294.

608. OSTLUND, E., 1954. The distribution of catechol amines in lower animals and their effects on the heart. *Acta Physiol. Scand.* **31,** Suppl. 112, 67.

609. PAINE, V. L., 1926. Adhesion of the tube feet in starfishes. *J. exp. Zool.* **45,** 361-366.

610. PAINE, V. L., 1929. The tube feet of starfishes as autonomous organs. *Am. Nat.* **63,** 517.

611. PAJETTA, E. and SAMUELLI, C., 1961. Richerche sulla affinita delle proteine degli Echinodermi. *Pubbl. Staz. zool. Napoli.* **32,** 1-8.

612. PANTIN, C. F. A., 1935. The nerve net of the Actinozoa. *J. exp. Biol.* **12,** 119-138.

613. PANTIN, C. F. A. and SAWAYA, P., 1953. Muscular action in *Holothuria grisea. Zoologia, Univ. S. Paulo* **18,** 51-59.

614. PARKER, B. and COLE, W. H., 1940. Studies of the body fluids and sera of some marine invertebrates. *Bull. Mt. Desert Is. biol. Lab.* 1939, 36-38.

615. PARKER, G. H., 1921. The locomotion of the holothurian *Stichopus parvimensis. J. exp. Zool.* **33,** 205-208.

616. PEARSE, A. S., 1908. Observations on the behaviour of the holothurian *Thyone briareus* (Leseur). *Biol. Bull. mar. biol. Lab. Woods Hole* **15,** 259-288.

617. PEARSE, J. S., 1965. Reproductive periodicities in several contrasting populations of *Odontaster validus* Koller, a common Antarctic asteroid. *Antarctic Res.*, Ser. 5, 1965, 39-85.

618. PEARSE, J. S., 1966. Antarctic asteroid *Odontaster validus*: Constancy of reproductive periodicities. *Science, N.Y.* **152,** 1763-1764.

619. PEARSE, J. S., 1967. Coelomic water volume control in the Antarctic sea star *Odontaster validus. Nature, Lond.* **216,** 1118-1119.

620. PEARSE, J. S. and GIESE, A. C., 1966a. Food, reproduction and organic constitution of the common Antarctic echinoid *Sterechinus neumayeri. Biol. Bull. mar. biol. Lab. Woods Hole* **130,** 387-401.

621. PEARSE, J. S. and GIESE, A. C., 1966b. The organic constituents of several benthonic invertebrates from McMurdo Sound, Antarctica. *Comp. Biochem. Physiol.* **18,** 47-57.

622. PEQUIGNAT, E., 1966. "Skin digestion" and epidermal absorption in irregular and regular urchins and their probable relation to the outflow of spherule-coelomocytes. *Nature, Lond.* **210,** 397-399.

623. PÉRÈS, J.-M., 1949. Recherches sur les pédicellaires glandulaires de *Sphaerechinus granularis. Archs. Zool. exp. gén.* **86,** 118-136.

624. PESKIN, J. C., 1951. Photolabile pigments in Invertebrates. *Science, N.Y.* **114,** 120-121.

625. PHILPOTT, D. E. and CHAET, A. B., 1960. Electron microscope observations of the starfish eyespot. *Biol. Bull. mar. biol. Lab. Woods Hole* **119,** 332-333.

626. PHILPOTT, D. E., CHAET, A. B. and BURNETT, A. L., 1966. A study of the secretory granules of the basal disc of hydra. *J. Ultrastruct. Res.* **14,** 74-84.

627. PICKEN, L. E. R., 1936. Mechanism of urine formation in certain invertebrates. I. Arthropoda. *J. exp. Biol.* **13,** 309-328.

628. PILKEY, O. H. and HOWER, J., 1960. The effect of environment on the concentrations of skeletal magnesium and strontium in *Dendraster. J. Geol.* **68,** 203-216.

629. PINCUS, G., THIMANN, K. V. and EASTWOOD, E. B. (Eds.), 1964. *The Hormones.* Academic Press, New York, 688 pp.

630. PLESSNER, H., 1913. Untersuchungen uber die Physiologie der Seesternes. I. Mitteilung Der Lichsinn. *Zool. Jb.* **33,** 361-386.

631. POLUNINA, O. A., 1965. On the capacity of the starfish *Asterias rubens* to differentiate the salinity of sea water. *Dokl. Akad. Nauk. SSSR* **165,** 451-453.

632. POPE, E., 1964. A stinging by a crown-of-thorns starfish. *Aust. Nat. Hist.* **14,** 350.

633. POPLE, W. and EWER, D. W., 1954. Studies on the myoneural physiology of Echinodermata. I. The pharyngeal retractor muscle of *Cucumaria sykion. J. exp. Biol.* **31,** 114-126.

634. POPLE, W. and EWER, D. W., 1955. Studies on the moyneural physiology of the Echinodermata. II. Circumoral conduction in *Cucumaria. J. exp. Biol.* **32,** 59-69.

635. POPLE, W. and EWER, D. W., 1958. Studies on the myoneural physiology of Echinodermata. III. Spontaneous activity of the pharyngeal retractor muscle of *Cucumaria. J. exp. Biol.* **35,** 712-730.

636. PORA, E. A., 1936. Sur les differences chimiques et physico-chimiques du sang des deux sexes de quelques invertebrès marins. *Bull. Inst. oceanogr. Monaco* **689,** 1-4.

637. POTTS, W. T. W., 1958. The inorganic and amino acid composition of some lamellibranch muscles. *J. exp. Biol.* **35,** 749-764.

638. Powell, V. H. and Sutherland, M. D., 1967. Pigments of marine animals. VI. Anthraquinoid pigments of the crinoids *Ptilometra australis* and *Tropiometra afra*, *Aust. J. Chem.* **20,** 541-553.

639. Powell, V. H., Sutherland, M. D. and Wells, J. W., 1967. Pigments of marine animals. V. Rubrocomatulin monomethyl ether, an anthraquinoid pigment of the *Comatula* genus of crinoids. *Aust. J. Chem.* **20,** 535-540.

640. Preyer, W., 1886. Uber die Bewegungen der Seesterne. *Mitt. zool. Sta. Neapel* **7,** 191-233.

641. Prosser, C. L., 1954. Activation of a non-propagating muscle in *Thyone. J. cell. comp. Physiol.* **44,** 247-253.

642. Prosser, C. L. and Brown, F. A., 1961. *Comparative Animal Physiology*, 2nd edn. Saunders, Philadelphia, 688 pp.

643. Prosser, C. L., Curtis, H. J. and Travis, D. M., 1951. Action potentials from some invertebrate nonstriated muscles. *J. cell. comp. Physiol.* **38,** 299-319.

644. Prosser, C. L. and Judson, C. L., 1952. Pharmacology of haemal vessels of *Stichopus californicus*. *Biol. Bull. mar. biol. Lab. Woods Hole* **102,** 249-251.

645. Prosser, C. L., Nystrom, R. A. and Nagai, N., 1965. Electrical and mechanical activity in intestinal muscles of several invertebrate animals. *Comp. Biochem. Physiol.* **14,** 53-70.

646a. Prouho, H., 1887. Recherches sur le *Dorocidaris papillata*. *Arch. Zool. exp. gen.* **5,** 213-380.

646b. Prouho, H., 1890. Du role des pedicellaires gemmiformes des Oursins. *C. r. hebd. Séanc. Acad. Sci. Paris* **111,** 62-64.

647. Quin, L. D., 1965. The presence of compounds with a carbon-phosphorus bond in some marine invertebrates. *Biochemistry, N.Y.* **4,** 324-330.

648. Quinn, B. G., 1965. Predation in sea urchins. *Bull. mar. Sci. Gulf Caribb.* **15,** 259-264.

649. Quinton, R., 1899. Communication osmotique: Chez l'invertebrès marin normal entre le milieu interne de l'animal et le milieu exterieux. *C. r. hebd. Séanc. Acad. Sci. Paris* **131,** 905-908; 952-955.

650. Randall, J. E., 1965. Grazing effect on sea grasses by herbivorous reef fishes in the West Indies. *Ecology* **46,** 255-260.

651. Randall, J. E., Schroeder, R. E. and Starch, W. A., II., 1964. Notes on the biology of the echinoid *Diadema antillarum*. *Carib. J. Sci.* **4,** 421-433.

652. Rao, K. S., 1965. Reproductive cycle of *Oreaster* (*Pentaceros*) *hedemanni* in relation to chemical composition of the gonads. *Curr. Sci.* **34,** 87-88.

653. Raup, D. M., 1959. Crystallography of echinoid calcite. *J. Geol.* **67,** 661-674.

654. Raup, D. M., 1960. Ontogenetic variation in the crystallography of echinoid calcite. *J. Palaeont.* **34,** 1041-1050.

655. Raup, D. M., 1966a. Crystallographic data for echinoid coronal plates. *J. Palaeont.* **40,** 555-568.

656. Raup, D. M., 1966b. The Endoskeleton, in *Physiology of Echinodermata*, Ed. R. A. Boolootian, John Wiley, New York.

657. Reese, A. M., 1942. The old starfish-clam question. *Science, N.Y.* **96,** 513-515.

658. REICHENSPERGER, A., 1908. Uber das Vorkommen von Drüsen bei Crinoiden. *Zool. Anz.* **33,** 363-367.

659. REISSER, O., 1931. Beitrage zur Kenntnis des Azetylocholins. *Arch. exp. Path. Pharmack.* **161,** 34-58.

660. REISSER, O., 1933. Fortgesetzte pharmacologische Untersuchungen an den Muskeln wirbelloser Meerestiere. *Arch. exp. Path. Pharmack.* **172,** 194-212.

661. RIO, G. J., RUGGIERI, G. D., MARTIN, S. J., STAMPIEN, F., JR. and NIGRELLI, R. F., 1963. Saponin-like toxin from the giant sunburst starfish *Pycnopodia helianthoides* from the Pacific north west. *Am. Zool.* **3,** 554-555.

662. RIO, G. J., STAMPIEN, F., JR., NIGRELLI, R. F. and RUGGIERI, G. D., 1965. Echinoderm toxins. I. Some biochemical and physiological properties of toxins from several species of Asteroidea. *Toxicon* **3,** 147-155.

663. ROAF, H. E., 1909. Contributions to the physiology of marine invertebrates. *J. Physiol., Lond.* **39,** 438.

664. ROBERTSON, J. D., 1939. Ionic composition of the bloods of *Homarus, Cancer* and *Echinus. J. exp. Biol.* **16,** 387-397.

665. ROBERTSON, J. D., 1949. Ionic regulation in some marine invertebrates. *J. exp. Biol.* **26,** 182-200.

666. ROBERTSON, J. D., 1953. Further studies on ionic regulation in marine invertebrates. *J. exp. Biol.* **30,** 277-296.

667. ROCKSTEIN, M., 1956. Role of the terminal pigment spots of the starfish *Asterias forbesi* in light orientation. *Nature, Lond.* **177,** 341-342.

668. ROCKSTEIN, M., 1962. Some properties of stellarin, the photosensitive pigment of the starfish *Asterias forbesi. Biol. Bull. mar. biol. Lab. Woods Hole* **123,** 510.

669. ROCKSTEIN, M., COHEN, J. and HAUSMANN, S. A., 1958. A photosensitive pigment from the dorsal skin and eyespots of the starfish *Asterias forbesi. Biol. Bull. mar. biol. Lab. Woods Hole* **115,** 361.

670. ROCKSTEIN, M. and FINKEL, A., 1960. Stellarin: a photosensitive pigment from the dorsal skin of the starfish *Asterias forbesi. Anat. Rec.* **138,** 379.

671. ROCKSTEIN, M. and RUBENSTEIN, M., 1957. The biochemical basis for positive photokinesis of the starfish *Asterias forbesi. Biol. Bull. mar. biol. Lab. Woods Hole* **113,** 353-354.

672. ROCKSTEIN, M. and SPRITZER, R., 1960. Light orientation in the starfish *Asterias forbesi. Anat. Rec.* **138,** 379.

673. RODEGKER, W. and NEVENZEL, J. C., 1964. The fatty acid composition of three marine invertebrates. *Comp. Biochem. Physiol.* **11,** 53-60.

674. RODENHOUSE, I. Z. and GUBERLET, J. E., 1946. The morphology and behaviour of the cushion star *Pteraster tesselatus. Univ. Wash. Publs. Biol.* **12,** 21-48.

675. ROMANES, G. J., 1884. Observations on the physiology of Echinodermata. *J. Linn. Soc.* **17,** 131-137.

676. ROMANES, G. J. and EWART, J. C., 1881. Observations on the locomotor system of Echinodermata. *Phil. Trans. R. Soc.* **172,** 829-885.

677. ROSE, B. and BROWNE, J. S. L., 1942. Studies on the blood histamine in cases of burns. *Ann. Surg.* **115,** 390-399.

678. ROUSHDY, H. M. and HANSEN, V. K., 1960. Ophiuroids feeding on phytoplankton. *Nature, Lond.* **188,** 517-518.

679. RUGGIERI, G. D., 1965. Echinoderm toxins. II. Animalizing action in sea urchin development. *Toxicon* **3,** 157-162.

680. RUGGIERI, G. D. and NIGRELLI, R. F., 1960. The effects of holothurin, a steroid saponin from the sea cucumber, on the development of the sea urchin. *Zoologica, N.Y.* **45,** 1-16.

681. RUGGIERI, G. D. and NIGRELLI, R. F., 1964. Effects of extracts of the red web starfish *Patiria miniata* on sea urchin eggs. *Am. Zool.* **4,** 431.

682. RUNDLES, C. and FARMANFARMAIAN, A., 1964. Absorption and transport of D-glucose in the intestine of *Thyone briareus. Biol. Bull. mar. biol. Lab. Woods Hole* **127,** 387-388.

683. RUSSELL, A. P., PATT, D. I. and TERNER, C., 1964. Invertebrate acid deoxyribonucleases. *J. cell. comp. Physiol.* **63,** 71-75.

684. RUSSELL, E. S., 1919. Notes on the righting reaction in *Asterina gibbosa. Proc. zool. Soc. Lond.*, pp. 423-432.

685. SAGARA, J. and INO, T., 1954. Optimum temperature and specific gravity for the bipinnaria and the young of the Japanese star *Asterias amurensis. Bull. Jap. Soc. scient. Fish.* **20,** 689-693.

686. SANDEMAN, D. C., 1965. Electrical activity in the radial nerve cord and ampullae of sea urchins. *J. exp. Biol.* **43,** 247-256.

687. SANDEMAN, E. J., 1954. Experimentally induced spawning in *Asterias rubens. Rep. Challenger Soc.* **3,** 14.

688. SANZO, L., 1907. Zur Kenntnis des Stickstoff-Stoffwechsels bei marinen wirbellosen Tieren. *Biol. Zbl.* **27,** 479-491.

689. SARCH, M. N., 1931. Die Pufferung der Korperflussigkeiten der Echinodermen. *Z. vergl. Physiol.* **14,** 525-545.

690. SASTRY, A. N., 1966. Variation in reproduction of latitudinally separated populations of two marine invertebrates. *Am. Zool.* **6,** 325-326.

691. SAWANO, E., 1928. On the digestive enzymes of *Caudina chilensis. Sci. Rep. Tohoku Imp. Univ.*, Ser. 4. **3,** 203-218.

692. SAWANO, E., 1936. Contributions to the knowledge of the digestive enzymes in invertebrates. II. Proteolytic enzymes in the starfish *Distolasterias nipon. Sci. Rep. Tokyo Bunrika Daig.*, Sect. B. **2,** 179-199.

693. SAWANO, E. and MITSUGI, K., 1932. Toxic effects of the stomach extracts of the starfishes on the heart of the oyster. *Sci. Rep. Tohoku Imp. Univ.*, Ser. 4. **7,** 79-88.

694. SAWAYA, P., 1951. Sensibilidadae do muscolo longitudinal radial de Holothuria à Acetilcolina. *Ciênc. e Cult.* **3,** 41-42.

695. SAWAYA, P., 1962. On a bioassay for acetylcholine and on some properties of the longitudinal muscles of *Holothuria grisea. Bol. Fac. Filos. Ciênc. S. Paulo* (*Zool.*) **24,** 5-10.

696. SAWAYA, P. and ANCONA-LOPEZ, A. A., 1959. Sobre a Fisiologia dos musculos longitudinais de *Holothuria grisea. Zoologia* **22,** 75-97.

697. SCHAEFER, M. B., 1964. *An Investigation of the Effects of Discharged Wastes on Kelp.* State Water Quality Control Board, Sacramento, Calif. 124 pp.

698. SCHAFER, E. A., 1883. Preliminary notice of an investigation into the coagulation of the perivisceral fluid of the sea urchin. *Proc. R. Soc.* **34,** 370-371.

699. SCHEER, B. T., 1940. Some features of the metabolism of the carotenoid pigments of the California sea mussel, *Mytilus californianus. J. Biol. Chem.* **136,** 275-299.

700. SCHIEMENZ, P., 1895-97. How do starfishes open oysters? *J. mar. biol. Ass. U.K.* **4,** 266-285.

701. Schinke, H., 1950. Bildung und ersatz zellelement der Leibesholhlen-flussigkeit von *Psammechinus miliaris*. *Z. Zellforsch. mikrosk. Anat.* **35,** 311-331.

702. Schlieper, C., 1929. Uber die Einwerkung niederer Saltzkonzentrationen auf marine Organismen. *Z. vergl. Physiol.* **9,** 478-514.

703. Schlieper, C., 1957. Comparative study of *Asterias rubens* and *Mytilus edulis* from the North Sea and the western Baltic Sea. *Ann. Biol.* **33,** 117-127.

704. Schreiber, B., 1930. Studi sull assorbimento intestinale nelle Oloturie. *Pubbl. Staz. zool. Napoli* **10,** 235-277.

705. Schreiber, B., 1932. Esperimenti per lo studio dell' assorbimento intestinale nelle Oloturie. *Archo zool. ital.* **16,** 865-870.

706. Schultz, E., 1895. Ueber den Prozess der Exkretion bei den Holothurien. *Biol. Zbl.* **15,** 390-398.

707. Seck, C., 1957. Untersuchungen zur Frage der Ionenregulation bei in Brackwasser lebenden Evertebraten. *Kieler Meeresforsch.* **13,** 220-243.

708. Segerstrale, S., 1949. Brackish water fauna of Finland. *Oikos* **1,** 127-141.

709. Serra von Buddenbrock, E., 1963. Études physiologiques et Histologiques sur le tégument les holothuries (*Holothuria tubulosa*). *Vie Milieu* **14,** 55-70.

710. Sharp, D. T. and Gray, I. E., 1962. Studies on factors affecting the local distribution of two sea urchins *Arbacia punctulata* and *Lytechinus variegatus*. *Ecology* **43,** 309-313.

711. Shlyakhter, T. A., 1959. Comparative study of heat resistance of ciliated epithelium cells of some species of starfish. *Tsitologyia* **1,** 369-373.

712. Simon, S. E., Edwards, S. and Dewhurst, D. J., 1964. Potassium exchange in holothuroidean muscle. *J. cell. comp. Physiol.* **63,** 89-100.

713. Simon, S. E., Muller, J. and Dewhurst, D. J., 1964. Ionic partition in holothuroidean muscle. *J. cell. comp. Physiol.* **63,** 77-84.

714. Simpson, J. W., Allen, K. and Awapara, J., 1959. Free amino acids in some aquatic invertebrates. *Biol. Bull. mar. biol. Lab. Woods Hole* **117,** 371-381.

715. Smith, E. H., 1962. Studies of *Cucumaria curata* Cowles 1907. *Pacific Nat.* **3,** 233-246.

716. Smith, G. F. M., 1940. Factors limiting distribution and size in the starfish. *J. Fish. Res. Bd. Can.* **5,** 84-103.

717. Smith, J. E., 1937. On the nervous system of *Marthasterias glacialis*. *Phil. Trans. R. Soc.*, B. **227,** 111-173.

718. Smith, J. E., 1938. The structure and function of the tube feet in certain echinoderms. *J. mar. biol. Ass. U.K.* **22,** 345-357.

719. Smith, J. E., 1945. The role of the nervous system in some activities of starfish. *Biol. Rev.* **20,** 29-43.

720. Smith, J. E., 1946. The mechanics and innervation of the starfish tube foot/ampulla systems. *Phil. Trans. R. Soc.*, B. **232,** 279-310.

721. Smith, J. E., 1947. The activities of the tube feet of *Asterias rubens*. *Q. Jl Microsc. Sci.* **88,** 1-14.

722. Smith, J. E., 1950a. The motor nervous system of *Astropecten irregularis*. *Phil. Trans. R. Soc.*, B. **234,** 521-558.

723. Smith, J. E., 1950b. Some observations on the nervous mechanisms underlying the behaviour of starfish. *Symp. Soc. exp. Biol.* **4,** 196-220.

724. Smith, J. E., 1965. Echinodermata, in *Structure and Function in the Nervous*

Systems of Invertebrates, by T. H. Bullock and G. A. Horridge. Freeman, San Francisco.

725. SMITH, J. E., 1966. The form and functions of the nervous system, in *Physiology of Echinodermata*, Ed. R. A. Boolootian. John Wiley & Sons, New York.

726. SMITH, R. I., 1955. Salinity variation in interstitial water of sand at Kames Bay, Millport, with reference to the distribution of *Nereis diversicolor*. *J. mar. biol. Ass. U.K.* **34,** 33-46.

727. SOKOLOV, V. A., 1961. Tactile conditioned reflex in the starfish *Asterias rubens*. *Trans. Murmansk Morskogo Biol. Inst.* **3,** 49-54.

728. SOKOLOV, V. A. and ASTAF'EVA, L. A., 1961. Destruction of stomach tissue in the starfish *Asterias rubens* L. as a response to the conditions under which it is kept. *Trans. Murmansk Morskogo Biol. Inst.* **3,** 55-60.

729. SOKOLOVA, M. N. and KUZNETSOV, A. P., 1960. On the feeding character and on the role played by trophic factors in the distribution of the hedgehog *Echinarachnius parma* (Lam). *Zool. Zh.* **39,** 1253-1256.

730. DESOUSA SANTOS, H., 1966. The ultrastructure of the mucus granules from the starfish tube feet. *J. Ultrastruct. Res.* **16,** 259-268.

731. SPARCK, R., 1932. On the capability of migration of adult individuals of *Asterias rubens*. *Rep. Dan. biol. Stn.* **37,** 65-68.

732. STEEN, J. B., 1965. Comparative aspects of the respiratory gas exchange of sea urchins. *Acta Physiol. Scand.* **63,** 164-170.

733. STEINBACH, H. B., 1937. Potassium and chloride in *Thyone* muscle. *J. cell. comp. Physiol.* **9,** 429-435.

734. STEINBACH, H. B., 1940. Electrolytes in *Thyone* muscles. *J. cell. comp. Physiol.* **15,** 1-9.

735. STEPHENS, G. C. and SCHINSKE, R. A., 1961. Uptake of amino acids by marine invertebrates. *Limnol. Oceanogr.* **6,** 175-181.

736. STEPHENS, G. C. and VIRKAR, R. A., 1966. Uptake of organic material by aquatic invertebrates. IV. The influence of salinity on the uptake of amino acids by the brittle star *Ophiactis arenosa*. *Biol. Bull. mar. biol. Lab. Woods Hole* **131,** 172-185.

737. STEPHENSON, R. A. and UFRET, S. L., 1966. Iron, manganese, and nickel in skeletons and food of the sea urchins *Tripneustes esculentus* and *Echinometra lacunter*. *Limnol. Oceanogr.* **11,** 11-17.

738. STONE, E. A., 1897. Some observations on the physiological function of the pyloric caeca of *Asterias rubens*. *Am. Nat.* **31,** 1035-1041.

739. STOTT, F. C., 1931. The spawning of *Echinus esculentus*, and some changes in gonad composition. *J. exp. Biol.* **8,** 133-150.

740. STOTT, F. C., 1955. The food canal of the sea urchin *Echinus esculentus* and its functions. *Proc. Zool. Soc. Lond.* **125,** 63-86.

741. STOTT, F. C., 1957. Observations on the food canal and associated structures in the holothurian *Holothuria forskali*. *Proc. zool. Soc. Lond.* **129,** 129-136.

742. SULLIVAN, T. D., LADUE, K. T. and NIGRELLI, R. F., 1955. The effects of holothurin, a steroid saponin of animal origin, on Krebs-2 ascites tumors in Swiss mice. *Zoologica, N.Y.* **40,** 49-52.

743. SULLIVAN, T. D. and NIGRELLI, R. F., 1956. The antitumerous action of biologics of marine origin. I. Survival of Swiss mice inoculated with Krebs-2 ascites tumour and treated with holothurin, a steroid saponin from the sea cucumber *Actinopyga agassizi*. *Proc. Am. Ass. Cancer Res.* **2,** 151.

744. SUTHERLAND, M. D. and WELLS, J. W., 1959. Anthraquinone pigments from the crinoid *Comatula pectinata*. *Chem. Ind.* **78,** 291-292.

745. SUTHERLAND, M. D. and WELLS, J. W., 1967. Pigments of marine animals. IV. The anthraquinoid pigments of the crinoids *Comatula pectinata* and *C. cratera*. *Aust. J. Chem.* **20,** 515-533.

746. SVEDBERG, T. and PETERSEN, K. O., 1940. *The Ultracentrifuge*. Oxford Univ. Press.

747. TADOKORO, T. and WATANABE, S., 1928. Chemical studies on sex differences of blood protein in *Caudina chilensis*. *Sci. Rep. Tohoku Imp. Univ.*, Ser. 4. **3,** 535-545.

748. TAKAHASHI, K., 1964. Electrical responses to light stimuli in the isolated radial nerve of the sea urchin *Diadema setosum*. *Nature, Lond.* **201,** 1343-1344.

749. TANAKA, Y., 1958. Feeding and digestive processes of *Stichopus japonicus*. *Bull. Fac. Fish. Hokkaido Univ.* **9,** 14-28.

750. TAO, L., 1927. Physiological characteristics of *Caudina* muscle with some accounts on the innervation. *Sci. Rep. Tohoku Imp. Univ.*, Ser. 4. **2,** 265-291.

751. TAO, L., 1930. Notes on the ecology and the physiology of *Caudina chilensis* (Muller) in Matsu Bay. *Proc. 4th Pacific Sci. Congr. 1929* **3,** 7-11.

752. THEEL, J. H., 1896. Remarks on the activity of amoeboid cells in the echinoderms. *Fest. Lilljeborg*, **1896,** pp. 47-58.

753. THEEL, J. H., 1921. On amoebocytes and other coelomic corpuscles in the perivisceral cavity of echinoderms. III. Holothurids. *Ark. Zool.* **13,** no. 25, 1-40.

753b. THOAI, N. V., ROCHE, J., ROBIN, Y. and THIEM, N. V., 1953. Sur la presence de la glycocyamine (Acid guanidyl acetique) de la taurocyamine (guanidyl-taurine) et des phosphagenes correspondants dans les muscles de vers marins. *Biochim. biophys. Acta* **11,** 593.

754. THOMAS, L. P., 1961. Distribution and salinity tolerance in the amphiurid brittle star *Ophiophragmus filograneous*. *Bull. mar. sci. Gulf Caribb.* **11,** 158-160.

755. THOMAS, L. P. and THOMAS, S. B., 1965. Herring gulls diving for starfish. *Q. Jl Florida Acad. Sci.* **28,** 195-196.

756. THORNTON, I. W. B., 1956. Diurnal migrations of the echinoid *Diadema setosum*. *Brit. J. Anim. Behav.* **4,** 143-146.

757. THORSON, G., 1946. Reproduction and larval development of the Danish marine bottom invertebrates, with special reference to the planktonic larvae of the Sound (Oresund). *Meddr Kommn Havunders. Ser. Plankt.* **4,** No. 1, 523 pp.

758. THRON, C. D., 1964. Haemolysis by Holothurin A, digitonin and Quillaia saponin: Estimates of the required cellular lysin uptake and free lysin concentrations. *J. Pharmac. exp. Ther.* **145,** 194-202.

759. THRON, C. D., DURRANT, R. C. and FRIES, S. L., 1964. Neuromuscular and cytotoxic effects of Holothurin A and related saponins at low concentration levels. *Toxicol. App. Pharmac.* **6,** 182-196.

760. TODD, D. and RIZZI, G. P., 1964. Biochemistry of the α-glyceryl ethers. I. Distribution in mammalian tissues and starfish. *Proc. Soc. exp. Biol. Med.* **115,** 218-222.

761. TOPPING, F. L. and FULLER, J. F., 1942. Osmotic concentration as related to distribution of marine and brackish invertebrates. *Biol. Bull. mar. biol. Lab. Woods Hole* **82,** 372-384.

762. TORTONESE, E. and DEMIR, M., 1960. The echinoderm fauna of the Sea of Marmara and the Bosphorous. *Hidrobiologi* **5B,** 3-16.

763. TOYAMA, Y. and TAKAGI, T., 1954a. Sterols and other unsaponifiable substances in the lipids of shell fish, crustacea and echinoderms. XIII. Sterol components of the starfish *Luidia quinaria. Bull. Chem. Soc. Japan* **27,** 39-41.

764. TOYAMA, Y. and TAKAGI, T., 1954b. Sterols and other unsaponifiable substances in the lipids of shell fishes, crustacea and echinoderms. XV. Occurrence of Δ^7 cholestenol as a sterol component of starfish *Asterias amurensis. Bull. Chem. Soc. Japan* **27,** 421-423.

765. TOYAMA, Y. and TAKAGI, T., 1955. Sterols and other unsaponifiable substances in the lipids of shell fishes, crustacea and echinoderms. XVI. Re-investigation of hitodestrol and its identity with α-spinasterol. *Bull. Chem. Soc. Japan* **28,** 469-473.

766. TOYAMA, Y. and TAKAGI, T., 1956. Sterols and other unsaponifiable substances in the lipids of shell fishes, crustacea and echinoderms. XVII. Mono-unsaturated sterol components of the starfish *Asterina pectinifera. Bull. Chem. Soc. Japan* **29,** 317-319.

767. TOYAMA, Y., TAKAGI, T. and TANAKA, T., 1955. Fatty oils of aquatic invertebrates. *Mem. Fac. Eng. Nagoya Univ.* **7,** 1-35.

768. TREFZ, S. M., 1956. Observations on the anatomy, both macroscopical and microscopical, of the intestinal tract and of accompanying organs and on the physiology of digestion in *Holothuria atra* (Jager) with additional observations on the ecology of holothurians and their possible role in the destruction of coral reefs. U.S.A. Navy Dept. Office Naval Res.

769. TSURNAMAL, M. and MARDER, J., 1966. Observations on the basket star *Astroboa nuda* (Lyman) on coral reefs at Elat (Gulf of Aqaba). *Israel J. Zool.* **15,** 9-17.

770. TSUMAKI, T., YAMAGUCHI, M., KAWASAKI, H. and MUKAI, T., 1954. Pigments of marine animals. II. The red and blue pigments of starfish *Asterina pectinifera. J. Jap. Chem. Soc.* **75,** 605.

771. TURSCH, B., DESOUSA GUIMARAES, I. S., GILVERT, B., APLIN, R. T., DUFFIELD, A. M. and DJERASSI, C., 1967. Chemical studies of marine invertebrates. II. Terpenoids. LVIII, Griseogenin, a new triterpenoid sapogenin of the sea cucumber *Halodeima grisea* L. *Tetrahedron* **23,** 761-767.

772. TYLER, A., 1939. Crystalline echinochrome and spinochrome: their failure to stimulate the respiration of eggs and sperm of *Strongylocentrotus. Proc. natl. Acad. Sci. Wash.* **25,** 523-528.

773. TYLER, A., 1946. Natural heteroagglutinins in the body fluids and seminal fluids of various invertebrates. *Biol. Bull. mar. biol. Lab. Woods Hole* **90,** 213-219.

774. VON UEXKULL, J., 1896a. Uber reflex bei den Seeigeln. *Z. Biol.* **34,** 298-318.

775. VON UEXKULL, J., 1896b. Vergleichend sinnesphysiologische Untersuchungen. II. Der Schatten als Reiz fur *Centrostephanus longispinus. Z. Biol.* **34,** 319-339.

776. VON UEXKULL, J., 1896c. Uber die Funktion der Poli'schen Blasen am Kauapparat der regularen Seeigel. *Mitt. Zool. Staz. Neapel.* **12,** 463-476.

777. von Uexkull, J., 1900a. Die physiologie des Seeigelstachels. *Z. Biol.* **39,** 73-112.
778. von Uexkull, J., 1905. Studien uber den Tonus. II. Die Bewegungen der Schlangensterne. *Z. Biol.* **46,** 1-37.
779. Unger, H., 1960. Neurohormone bei Seesternen (*Marthasterias glacialis*). *Symp. Biol. Hungarica* **1,** 203-207.
780. Unger, H., 1962. Experimentelle und histologische Untersuchungen uber Wirkfaktoren aus dem Nervensystem von *Asterias rubens. Zool. Jb.* (*Physiol.*) **69,** 481-536.
781. Ursin, E., 1960. A quantitative investigation of the echinoderm fauna of the central North Sea. *Medd. Danmarks. Fish Havunders.* **2,** 1-204.
782. Van der Heyde, H. C., 1922a. *On the Physiology of Digestion, Respiration and Excretion in Echinoderms.* Der Boer, Den Helder, Holland. 111 pp.
783. Van der Heyde, H. C., 1922b. Hemoglobin in *Thyone briareus. Biol. Bull. mar. biol. Lab. Woods Hole* **42,** 95-98.
784. Van der Heyde, H. C., 1923. Sur l'excretion chez les echinodermes. *Archs neerl. Physiol.* **8,** 151-160.
785. Vasu, B. S. and Giese, A. C., 1966. Protein and non-protein nitrogen in the body fluid of *Pisaster ochraceous* (Echinodermata) in relation to nutrition and reproduction. *Comp. Biochem. Physiol.* **19,** 351-361.
786. Vaupel-von Harnack, H., 1963. Uber den Feinbau des Nervensystems des seesternes (*Asterias rubens* L.). III. Mitteilung die struktur der Augenpolster. *Z. Zellforsch. mikrosk. Anat.* **60,** 432-451.
787. Ven, C., 1922. Sur la formation d'habitudes chez les asteries. *Archs. neerl. Physiol.* **6,** 163-178.
788. Verchowskaja, I., 1931. Experimental studien uber das axial organ von Asteroidea. *Z. vergl. Physiol.* **14,** 405.
789. Verrill, A. E., 1914. On the shallow water starfishes of the North Pacific Coast from the Arctic Ocean to California. *Harriman Alaska,* Ser. 14, part 1.
790. Vevers, H. G., 1951a. The biology of *Asterias rubens* L. II. Parasitization of the gonads by the ciliate *Orchitophrya stellarum. J. mar. biol. Ass. U.K.* **29,** 619-624.
791. Vevers, H. G., 1951b. The biology of *Asterias rubens* L. III. Carotenoid pigments in the integument. *J. mar. biol. Ass. U.K.* **30,** 569-574.
792. Vevers, H. G., 1954. Observations on feeding in Echinoderms. *Rep. Challenger Soc.* **3,** 15-16.
793. Vevers, H. G., 1956. Observations on feeding mechanisms in some echinoderms. *Proc. zool. Soc. Lond.* **126,** 484-485.
793b. Vevers, H. G., 1963. Pigmentation of the echinoderms. *Proc. XVI Int. Congr. Zool.* Washington.
794. Vevers, H. G., 1967. The chemistry of the echinoid axial organ, in *Echinoderm Biology,* Ed. N. Millott, Symp. zool Soc. London, No. 20.
795. Vevers, H. G. and Millott, N., 1957. Carotenoid pigments in the integument of the starfish *Marthasterias glacialis. Proc. zool. Soc. Lond.* **129,** 75-80.
796. Villela, G. G., 1951. On the fluorescent pigment of *Holothuria grisea. Rev. Bras. Biol.* **11,** 33-36.
797. Vinogradov, A. P., 1953. Elementary chemical composition of marine organisms. *Mem. Sears Fdn. mar. Res.* **2.**

798. VINOGRADOVA, Z. A., 1964. Biochemical composition of certain invertebrates found in the Black Sea. *Nauk. Zap. Odes'koyi Biol. Sta.* **5,** 26-33.

799. VYAZOV, O. E. and MURASHOVA, A. I., 1961. Normal antibodies during regeneration in invertebrate animals. I. Normal antibodies against regenerating tissue in the ray of *Asterias rubens* L. *Bull. exp. Biol. Med.* (*USSR*) **51,** 226-228.

800. WALD, G., 1945. The chemical evolution of vision. *Harvey Lect. 1945-46*, Ser. 41, 117-160.

801. WALD, G., 1952. In *Modern Trends in Physiology and Biochemistry*. Ed. E. S. G. Barron, Academic Press, New York.

802. WALLENFELS, K. and GAUHE, A., 1943. Synthese von Echinochrom A. *Chem. Ber.* **76,** 325-327.

803. WATTS, D. C., 1965. Evolutionary implications of enzyme structure and function, in *Studies in Comparative Biochemistry*. Ed. K. A. Munday, Pergamon Press, Oxford.

804. VAN WEEL, P. B., 1935. Uber die Lichtemfindlichkeit der Ambulakralfusschen des Seesternes (*Asterias rubens*). *Archs. neerl. Zool.* **1,** 347-353.

805. WELLS, G. P., 1942. The action of potassium on echinoderm molluscan and crustacean muscle. *J. exp. Biol.* **18,** 213-222.

806. WELLS, H. W., 1961. The fauna of oyster beds with special reference to the salinity factor. *Ecol. Monogr.* **31,** 239-266.

807. WELLS, H. W., WELLS, M. J. and GRAY, I. E., 1961. Food of the sea star *Astropecten articulatus*. *Biol. Bull. mar. biol. Lab. Woods Hole* **120,** 265-271.

808. WELSH, J. H., 1954. Marine invertebrate preparations useful in the bioassay of acetylcholine and 5-hydroxytryptamine. *Nature, Lond.* **173,** 955-956.

809. WELSH, J. H., 1966. Neurohumours and neurosecretion, in *Physiology of Echinodermata*. Ed. R. A. Boolootian, John Wiley, New York.

810. WELSH, J. H. and MOORHEAD, M., 1960. The quantitative distribution of 5-hydroxytryptamine in the invertebrates, especially in their nervous systems. *J. Neurochem.* **6,** 146-169.

811. WINTERSTEIN, H., 1909. Uber die Atmung der Holothurien. *Arch. Fisiol.* **7,** 87-93.

812. WOLF, E., 1925. Physiologische untersuchungen uber das umdrehen der seesterne und schlangensterne. *Z. vergl. Physiol.* **3,** 209-224.

813. WOODLEY, J. D., 1967. Problems in the ophiuroid water vascular system. *Symp. zool. Soc. Lond.* **20,** 75-104.

814. WYMAN, L. C. and LUTZ, B. R., 1930. The action of adrenalin, and certain drugs on the isolated holothurian cloaca. *J. exp. Zool.* **57,** 441-454.

815. YAMANOUCHI, T., 1926. Some preliminary notes on the behaviour of the holothurian *Caudina chilensis*. *Sci. Rep. Tohoku Imp. Univ.*, Ser. 4. **2,** 85-91.

816. YAMANOUCHI, T., 1929a. Notes on the behaviour of the holothurian *Caudina chilensis*. *Sci. Rep. Tohoku Imp. Univ.*, Ser. 4. **4,** 75-116.

817. YAMANOUCHI, T., 1929b. Statistical study on *Caudina chilensis*. *Sci. Rep. Tohoku Imp. Univ.*, Ser. 4. **4,** 335-359.

818. YAMANOUCHI, T., 1929c. Reactions to centrifugal force in the holothurian *Caudina chilensis*. *Sci. Rep. Tohoku Imp. Univ.*, Ser. 4. **4,** 521-532.

819. YAMANOUCHI, T., 1929d. Effects of anions in chemical stimulation in the holothurian *Caudina chilensis*. *Sci. Rep. Tohoku Imp. Univ.*, Ser. 4. **4,** 603-615.

820. YAMANOUCHI, T., 1955. On the poisonous substance contained in holothurians. *Publ. Seto. mar. biol. Lab.* **4,** 183-205.
821. YAMANOUCHI, T., 1956. The daily activity rhythms of the holothurians in the coral reef of Palao Islands. *Publ. Seto. mar. biol. Lab.* **5,** 347-362.
822. YAMASU, T. and KAWAGUTI, S., 1955. Naphthoquinone pigments of sea urchins. *Biol. J. Okayama Univ.* **2,** 62-66.
823. YANAGISAWA, T., 1959a. Studies on echinoderm phosphagens. I. Occurrence and nature of phosphagens in sea urchin eggs and spermatozoa. *J. Fac. Sci. Tokyo Univ.* **8,** 473-479.
824. YANAGISAWA, T., 1959b. Studies on guanidine phosphoryltransferases of echinoderms. I. Occurrence in spermatozoa and eggs of sea urchins. *J. Fac. Sci. Tokyo Univ.* **8,** 481-486.
825. YANAGISAWA, T., 1959c. Studies on the guanidine phosphoryltransferases of echinoderms. II. Some properties of the guanidine phosphoryltransferases of sea urchin eggs and spermatozoa. *J. Fac. Sci. Tokyo Univ.* **8,** 487-498.
826. YANAGISAWA, T., 1961. Studies on echinoderm phosphagens. II. Changes in phosphagen contents of sea urchin gonads during their maturation. *J. Fac. Sci. Tokyo Univ.* **9,** 197-204.
827. YARNALL, J. L., 1964. The responses of *Tegula funebralis* to starfishes and predatory snails. *Veliger* **6,** (Suppl.), 56-58.
828. YASUMOTO, T., NAKAMURA, K. and HASHIMOTO, Y., 1967. A new saponin, Holothurin B, isolated from sea cucumber *Holothuria vagabunda* and *Holothuria lubrica. Agric. Biol. Chem.* **31,** 7-10.
829. YAZAKI, M., 1930. On the circulation of the perivisceral fluid in *Caudina chilensis. Sci. Rep. Tohoku Imp. Univ.*, Ser. 4. **5,** 403-414.
830. YOKOE, Y. and YASUMASU, I., 1964. The distribution of cellulase in invertebrates. *Comp. Biochem. Physiol.* **13,** 323-338.
831. YOSHIDA, M., 1956. On the light response of the chromatophores of the sea urchin *Diadema setosum* (Leske). *J. exp. Biol.* **33,** 119-123.
832. YOSHIDA, M., 1957a. Spectral sensitivity of chromatophores in *Diadema setosum. J. exp. Biol.* **34,** 222-225.
833. YOSHIDA, M., 1957b. Positive phototaxis in *Psammechinus microtuberculatus. Pubbl. Staz. zool. Napoli* **30,** 260-262.
834. YOSHIDA, M., 1959. Naphthoquinone pigments in *Psammechinus miliaris. J. mar. biol. Ass. U.K.* **38,** 455-460.
835. YOSHIDA, M., 1962. The effect of light on the shadow reaction of the sea urchin *Diadema setosum* (Leske). *J. exp. Biol.* **39,** 589-602.
836. YOSHIDA, M. and MILLOTT, N., 1960. The shadow reaction of *Diadema antillarum.* III. Re-examination of the spectral sensitivity. *J. exp. Biol.* **37,** 390-397.
837. YOSHIDA, M. and OHTSUKI, H., 1966. Compound ocellus of a starfish: its function. *Science, N.Y.* **153,** 197-198.
838. YUDKIN, W. H., 1954. Transphosphorylation in echinoderms. *J. cell comp. Physiol.* **44,** 507-518.
839. ZENKEVITCH, L., 1963. *Biology of the Seas of the U.S.S.R.* Allen & Unwin, London, 955 pp.

ADDENDUM

THE following literature has not been referred to in the text and was either unknown to the author at the time of writing the appropriate chapter or was published subsequently.

Chapter 1. Feeding and Digestion

AHEARN, G. A., 1968. A comparative study of P^{32} uptake by whole animals and isolated body regions of the sea cucumber *Holothuria atra*. *Biol. bull. mar. biol. Lab. Woods Hole* **134,** 367-381.

BRAUER, R. W. and JORDAN, M. R., 1970. Triggering of the stomach eversion reflex of *Acanthaster planci* by coral extracts. *Nature, Lond.* **228,** 344-346.

CAMACHO, Z., BROWN, J. R. and KITTO, G. B., 1970. Purification and properties of trypsin-like proteases from the starfish *Dermasterias imbricata*. *J. biol. Chem.* **245,** 3964-3972.

CHAN, V. G. and FONTAINE, A. R., 1971. Is there a β-cell homology in starfish? *Gen. comp. Endocrinol.* **16,** 183-191.

CHRISTENSEN, A. G., 1970. Feeding biology of the sea star *Astropecten irregularis* Pennant. *Ophelia* **8,** 1-134.

DOEZEMA, P. and PHILLIPS, J. H., 1970. Glycogen storage and synthesis in the gut of the purple sea urchin *Strongylocentrotus purpuratus*. *Comp. Biochem. Physiol.* **34,** 691-697.

FARMANFARMAIAN, A., 1969a. Intestinal absorption and transport in *Thyone*. I. Biological aspects. *Biol. bull. mar. biol. Lab. Woods Hole* **137,** 118-131.

FARMANFARMAIAN, A., 1969b. Intestinal absorption and transport in *Thyone*. II. Observations on sugar transport. *Biol. bull. mar. biol. Lab. Woods Hole* **137,** 132-145.

FAVOROV, V. V. and VASKOVSKY, V. E., 1971. Alginases in marine invertebrates. *Comp. Biochem. Physiol.* **38,** 689-696.

FERGUSON, J. C., 1967. An autoradiographic study of the utilization of free exogenous amino acids by starfishes. *Biol. bull. mar. biol. Lab. Woods Hole* **133,** 317-329.

FERGUSON, J. C., 1968. Transport of amino acids by starfish digestive glands. *Comp. Biochem. Physiol.* **24,** 921-931.

FERGUSON, J. C., 1970. An autoradiographic study of the translocation and utilization of amino acids by starfish. *Biol. bull. mar. biol. Lab. Woods Hole* **138,** 14-25.

FONTAINE, A. R. and CHIA, F.-S., 1968. An autoradiographic study of assimilation of dissolved organic molecules. *Science, N.Y.* **161,** 1153-1155.

KOZLOVSKAYA, E. P. and VASKOVSKY, V. E., 1970. A comparative study of proteinases of marine invertebrates. *Comp. Biochem. Physiol.* **34,** 137-142.

KRISHNAN, S. and KRISHNASWAMY, S., 1970. Studies on the transport of sugars in the holothurian *Holothuria scabra*. *Mar. biol. Berlin* **5,** 303-306.

Lawrence, J. M., 1967. Lipid reserves in the gut of three species of tropical sea urchins. *Caribb. J. Sci.* **7,** 65-68.

Lawrence, J. M., 1970. The effect of starvation on the lipid and carbohydrate levels of the gut of the tropical sea urchin *Echinometra mathaei* (deBlainville). *Pac. Sci.* **24,** 487-489.

Lawrence, J. M., 1972. Carbohydrate and lipid levels in the intestine of *Holothuria atra* (Echinodermata: Holothuroidea). *Pac. Sci.* **26,** 114-116.

Leighton, D. L., Jones, L. G. and North, W. J., 1966. Ecological relationships between giant kelp and sea urchins in Southern California. *Proc. 5th Int. Seaweed Symp.* 141-153.

MacKenzie, C. L., 1970. Feeding rates of starfish *Asterias forbesi* (Desor) at controlled water temperatures and during different seasons of the year. *U.S. Fish. & Wildlife Serv. Fish. Bull.* **68,** 67-72.

McPherson, B. F., 1968. Feeding and oxygen uptake of the tropical sea urchin *Eucidaris tribuloides* (Lamarck). *Biol. bull. mar. biol. Lab. Woods Hole* **135,** 308-320.

Mauzey, K. P., Birkeland, C. and Dayton, P. K., 1968. Feeding behaviour of asteroids and escape responses of their prey in the Puget Sound region. *Ecology* **49,** 603-619.

Paine, R. T. and Vadas, R. L., 1969. The effects of grazing by sea urchins, *Strongylocentrotus* spp. on benthic algal populations. *Limnol. Oceanogr.* **14,** 710-719.

Pentreath, R. J., 1969. The morphology of the gut and a qualitative review of digestive enzymes in some New Zealand ophiuroids. *J. Zool. Lond.* **159,** 413-423.

Pentreath, R. J., 1970. Feeding mechanisms and the functional morphology of podia and spines in some New Zealand ophiuroids (Echinodermata). *J. Zool. Lond.* **161,** 395-429.

Pequinat, C. E., 1970. Biologie des *Echinocardium cordatum* (Pennant) de la Baie de Seine: Nouvelles recherches sur la digestion et l'absorption cutanees chez les echinides et les stellerides. *Forma Fonctio* **2,** 121-168.

Rosati, F., 1970. The fine structure of the alimentary canal of holothurians. II. The uptake of ferritin and iron dextran. *Monit. Zool. Ital.* **4,** 107-113.

Shibaeva, V. I., Elyakova, L. A. and Ovodov, Y. S., 1970. Pectic substances of Zosteraceae. III. Zosterinase activity of some invertebrates. *Comp. Biochem. Physiol.* **36,** 183-187.

Wilson, S. and Falkmer, S., 1965. Starfish insulin. *Can. J. Biochem.* **43,** 1615-1624.

Winter, W. P. and Neurath, H., 1970. Purification and properties of a trypsin-like enzyme from the starfish *Evasterias troschelli. Biochemistry* **9,** 4673-4679.

Chapter 2. Excretion and the Role of Amoebocytes

Brown, A. C., 1967. The elimination of foreign particles injected into the coelom of the holothurian *Cucumaria stephensoni* D. John. *Zool. Afr.* **3,** 3-8.

Chien, P. K., Johnson, P. T., Holland, N. D. and Chapman, F. A., 1970. The coelomic elements of sea urchins (*Strongylocentrotus*). IV. Ultrastructure of the coelomocytes. *Protoplasma* **71,** 419-442.

Johnson, P. T., 1969a. The coelomic elements of sea urchins (*Strongylocentrotus*). I. The normal coelomocytes; their morphology and dynamics in hanging drops. *J. Invert Pathol.* **13,** 25-41.

JOHNSON, P. T., 1969b. The coelomic elements of sea urchins (*Strongylocentrotus*). II. Cytochemistry of the coelomocytes. *Histochemie, Berlin* **17,** 213-231.

JOHNSON, P. T., 1970. The coelomic elements of sea urchins (*Strongylocentrotus* and *Centrostephanus*). VI. Cellulose acetate membrane electrophoresis. *Comp. Biochem. Physiol.* **37,** 289-300.

JOHNSON, P. T. and CHAPMAN, F. A., 1971. Comparative studies on the *in vitro* response of bacteria to invertebrate body fluids. III. *Stichopus tremulus* (Sea cucumber) and *Dendraster excentricus* (Sand dollar). *J. Invert. Pathol.* **17,** 94-106.

PRENDERGAST, R. A. and SUZUKI, M., 1970. Invertebrate protein stimulating mediators of delayed hypersensitivity. *Nature, Lond.* **227,** 277-279.

Chapter 3. Salinity Tolerance and Osmoregulation

ANCONA-LOPEZ, A. A., 1965. Contribution to the ecology of the Holothuroidea of the coast of Sao Paulo State. *Ann. Acad. Brazil* **37,** (Suppl.), 171-174.

EMERSON, D. N., 1969. Influence of salinity on ammonia excretion rates and tissue constituents of euryhaline invertebrates. *Comp. Biochem. Physiol.* **29,** 1115-1133.

KHLEBOVICH, V. V. and LUKANIN, V. V., 1967. Survival of the spermatozoa of certain White Sea invertebrates in water of varying salinity and temperature. *Dokl. Akad. Nauk. SSSR* **176,** 460-462.

LANDENBERGER, D. E., 1969. The effects of exposure to air on Pacific starfish and its relationship to distribution. *Physiol. Zool.* **42,** 220-230.

LEIGHTON, D., NUSSBAUM, I. and MULFORD, S., 1967. Effects of waste discharge from Pt. Loma saline water conversion plant on intertidal marine life. *J. Water poll. contr. Fed.* **39,** 1190-1202.

PERTSOVA, N. M. and SAKHARVA, M. I., 1970. Osobennosti razvitiya zooplanktona v. pribrezhnykh raionakh Kandalakshoyo zaliva (Velikaya Salma), 1966, 1967. *Tr. Belomorsk. Biol. Sta. Moskgas Univ.* **3,** 22-33. Trans. from *Ref. Zh. Biol.* 1970, No. 9, U 64.

WEIHE, S. C. and GRAY, I. E., 1968. Observations on the biology of the sand dollar *Mellita quinquiesperforata* (Leske). *J. Elisha Mitchell scient. Soc.* **84,** 315-327.

WOLFF, W. J., 1968. The echinodermata of the estuarine region of the rivers Ahine, Meuse and Scheldt, with a list of species occurring in the coastal waters of the Netherlands. *Neth. J. Sea Res.* **4,** 59-85.

Chapter 4. Ionic Regulation and Chemical Composition

AKINO, T., SHIMOJO, T. and SASAKI, T., 1970. Studies on visceral lipids of starfish I. Lipid components of various organs. *Sapporo Med. J.* **37,** 182-192.

ALLEN, W. V., 1968. Fatty acid synthesis in the echinoderms *Asterias rubens, Echinus esculentus* and *Holothuria forskali. J. mar. biol. Ass. U.K.* **48,** 521-533.

BACCETTI, B., 1967. High resolutions on collagen of Echinodermata. *Monit. zool. Ital. N.S.* **1,** 201-206.

DONNAY, G. and PAWSON, D. L., 1969. X-ray diffraction studies of echinoderm plates. *Science, N.Y.* **166,** 1147-1150.

DYER, W. J., 1952. Amines in fish muscle. VI. Trimethylene oxide content of fish and marine invertebrates. *J. Fish. Res. Bd. Can.* **8,** 314-324.

FECHTER, H., 1972. Untersuchungen über den Wasserwechsel der Seeigel und seine Bedeutung fur Atmung und Exkretion. *Helg. Meeresforsch.* **23,** 80-99.

FLETCHER, G. L., 1971. Accumulation of yellow phosphorus by several marine invertebrates and seaweed. *J. Fish. Res. Bd. Can.* **28,** 793-796.

HEATFIELD, B. M., 1970. Calcification in echinoderms: Effects of temperature and Diamox on incorporation of Ca^{45} *in vitro* by regenerating spines of *Strongylocentrotus purpuratus. Biol. Bull. mar. biol. Lab. Woods Hole* **139,** 151-163.

HEATFIELD, B. M., 1971. Growth of the calcareous skeleton during regeneration of spines of the sea urchin *Strongylocentrotus purpuratus* (Stimpson): A light and scanning electron microscope study. *J. Morphol.* **134,** 57-90.

KOBAYASHI, S. and TAKI, J., 1969. Calcification in sea urchins: I. A tetracycline investigation of growth of the mature test in *Strongylocentrotus intermedius. Calcif. tiss Res.* **4,** 210-223.

LAUGA, J. and LECHAL, J., 1966. Étude comparée du milieu interieux de quatre epèces d'holothuries. *Vie et Milieu* **17,** 1013-1025.

LEWIS, R. W., 1967. Fatty acid composition of some marine animals from various depths. *J. Fish. Res. Bd. Can.* **24,** 1101-1115.

MAUCHLINE, J. and TEMPLETON, W. L., 1966. Strontium, calcium and barium in marine organisms from the Irish Sea. *J. Cons. perm. int. Explor. Mer. Perma* **30,** 161-170.

MEYER, D. L., 1971. The collagenous nature of problematical ligaments in Crinoids (Echinodermata). *Mar. Biol. Berlin* **9,** 235-241.

MINA, K., 1970. Studies on a freezing storage of sea urchins. I. On pungent components in a frozen sea urchin gonad. *Bull. Jap. Soc. scient. Fish.* **36,** 617-622.

NISSEN, H.-U., 1969. Crystal orientation and plate structure in echinoid skeletal units. *Science, N.Y.* **166,** 1150-1152.

OVODOV, YU. S., GORSHOVA, R. P., TOMSHICH, S. V. and ADAMENKO, M. N., 1969. Polysaccharides of some Japan Sea invertebrates. *Comp. Biochem. Physiol.* **29,** 1093-1098.

RILEY, J. P., 1970. The distribution of the major and some minor elements in marine animals. I. Echinoderms and Coelenterates. *J. mar. biol. Ass. U.K.* **50,** 721-730.

ROKAVEC, A. and CERKOVNIKOV, E., 1967. Prilog proučavanju Jadranskog mora. *Acta Pharm. Jugoslav.* **17,** 65-72.

SHIMIZU, M., 1970. Calcification in sea urchins. II. The seasonal changes of protein concentration and electrophoretic patterns of both proteins and mucopolysaccharides in the perivisceral fluid of a sea urchin. *Bull. Jap. Soc. scient. Fish.* **36,** 377-384.

SIMON, G. and ROUSER, G., 1969. Species variations in phospholipid distribution of organs. II. Heart and skeletal muscle. *Lipids* **4,** 607-614.

SPRINGHALL, J. A. and DINGLE, J. G., 1967. Growth and pathology of chicks fed bêche de mer meal. *Aust. vet. J.* **43,** 298-303.

TOWE, K. M., 1967. Echinoderm calcite: Single crystal or polycrystalline aggregate. *Science, N.Y.* **157,** 1048-1050.

WEBER, J., GREER, R., WOIGHT, B., WHITE, E. and ROY, R., 1969. Unusual strength properties of echinoderm calcite related to structure. *J. Ultrastruct. Res.* **26,** 355-366.

Weber, J. N. and Raup, D. M., 1968. Comparison of C^{13}/C^{12} and O^{18}/O^{16} in the skeletal calcite of recent and fossil echinoids. *J. Palaeontol.* **42,** 37-50.

Zhukova, I. G., Bogdanovskaya, T. A., Smirnova, G. P. and Kochetkov, N. K., 1970. Sialoglykolipidy morskoi zvezdy *Distolasterias nippon.* Vydelenie kharakteristika monomernogo sostava i sialovykh. (Sialoglycolipids of the starfish *Distolasterias nippon.* Isolation and characteristics and monomeric composition of sialic acids.) *Biokhimyia* **35,** 775-780.

Chapter 5. Biochemical Affinities

Badman, S. W. and Brookbank, J. W., 1970. Serological studies of two hybrid sea urchins. *Develop. Biol.* **21,** 243-256.

Dhanani, Z. and Kitto, G. B., 1970. Comparative studies of echinoderm aspartate amino transferases. *Int. J. Biochem.* **1,** 213-229.

Ellingboe, J. and Karnovsky, M. L., 1967. Origin of glycerol ethers: Biosynthesis from labelled acetate, stearic acid, stearaldehyde and stearyl alcohol. *J. biol. Chem.* **242,** 5693-5699.

Goad, L. J., Rubinstein, I. and Smith, A. G., 1972. The sterols of Echinoderms. *Proc. R. Soc.* B **180,** 223-246.

Gupta, K. C. and Scheuer, P. J., 1968. Echinoderm sterols. *Tetrahedron* **24,** 5831-5837.

Smith, A. G. and Goad, L. J., 1971. Sterol biosynthesis in the starfish *Asterias rubens* and *Henricia sanguinolenta. Biochem. J.* **123,** 671-672.

Snyder, F., 1969. The biosynthesis of alkyl glyceryl ethers by microsomal enzymes of digestive glands and gonads of the starfish *Asterias forbesi. Biochim. biophys. Acta* **187,** 302-306.

Yanagisawa, T., 1967. Studies upon echinoderm phosphagens. III. Changes in the content of creatine phosphate after stimulation of sperm motility. *Exptl. cell Res.* **46,** 348-354.

Chapter 6. Toxins and Immunology

Anisimov, M. N., Prokof'eva, N. G., Kuznetzova, T. A. and Peretolchin, N. V., 1971. Vliyanie nekotorykh triterpenovykh glikozodov na sintez belka v kul'ture kletokhostnogo mozga krys. (The effect of some triterpene glycosides on protein synthesis in tissue cultures of the bone marrow of rats.) *Izv. Akad. Nauk SSSR. Ser. Biol.* **1,** 137-140.

Brown, R., Almodovar, L. R., Bhatia, H. M. and Boyd, W. C., 1968. Blood group specific agglutinins in invertebrates. *J. Immunol.* **100,** 214-216.

Brusle, J., 1967. Homogreffes et heterogreffes réciproques du teguments et des gonades chez *Asterina gibbosa* Pennant et *Asterina panceri* Gasco (Echinodermata: Asteroidea). *Cah. Biol. mar.* **8,** 417-420.

Brusle, J., 1970. Auto-, homo- and hetero-transplantations de gonades chez *Asterina gibbosa* Pennant. *Archs. Zool. exp. gen.* **111,** 159-180.

Cannone, A. J., 1970. The anatomy and venom emitting mechanism of the globiferous pedicellariae of the sea urchin *Parechinus angulosus* (Leske) with notes on their behaviour. *Zool. Afr.* **5,** 179-190.

Clark, W. C., 1958. Escape responses of herbivorous gastropods when stimulated by carnivorous gastropods. *Nature, Lond.* **181,** 137-138.

CRUMP, R. G., 1968. The flight response in *Struthiolaria papulosa gigas* Sowerby. *N.Z. J. mar. freshwater Res.* **2,** 390-397.

EDWARDS, D. C., 1969. Predators on *Olivella biplicata* including a species specific predator avoidance response. *Veliger* **11,** 326-333.

FANGE, R., 1963. Toxic factors in starfishes. *Sarsia* **10,** 19-21.

FEDER, H. M., 1967. Organisms responsive to predatory sea stars. *Sarsia* **29,** 371-394.

FEDER, H. M. and ARVIDSSON, J., 1967. Studies on a sea star (*Marthasterias glacialis*) extract responsible for avoidance reactions in a gastropod (*Buccinum undatum*). *Arkiv. Zool.* **19,** 369-379.

FEIGEN, G. A., HADJI, L., PFEFFER, R. A. and MARKUS, G., 1970. Studies on the mode of attack of sea urchin toxin on natural and synthetic substrates. I. Conditions determining the formation and isolation of plasma kinins. *Physiol. Chem. Phys.* **2,** 309-322.

FEIGEN, G. A., SANZ, E., TOMITA, J. T. and ALENDER, C. B., 1968. Studies on the mode of action of sea urchin toxin. II. Enzymatic and immunological behaviour. *Toxicon* **6,** 17-43.

HABERMEHL, G. and VOLKWEIN, G., 1968. Ueber Gifte der mittelmeerischen Holothurien. (Poisons of Mediterranean holothurians.) *Naturwiss enschaften* **55,** 83-84.

HABERMEHL, G. and VOLKWEIN, G., 1970. Ueber Gifte der mittelmeerischen Holothurien. II. Die Aglyka der Toxine von *Holothuria poli.* (The aglycones of the toxin of *Holothuria poli.*) *Justus Liebigs Annln Chem.* **731,** 53-57.

HABERMEHL, G. and VOLKWEIN, G., 1971. Aglycones of the toxins from the Cuvierian organs of *Holothuria forskali* and a new nomenclature for the aglycones from Holothuroidea. *Toxicon* **9,** 319-326.

HILGARD, H. R., HINDS, W. E. and PHILLIPS, J. H., 1967. The specificity of uptake of foreign proteins by coelomocytes of the purple sea urchin. *Comp. Biochem. Physiol.* **23,** 815-824.

HILGARD, H. R. and PHILLIPS, J. H., 1968. Sea urchin response to foreign substances. *Science, N.Y.* **161,** 1243-1245.

HUET, M., 1967. Étude expérimentale du rôle système nerveux dans la régénération du bras l'étoile de mer *Asterina gibbosa* (Pennant)—échinoderme—asteride. *Bull. soc. Zool. Fr.* **92,** 641-645.

JOHNSON, P. T., 1969. The coelomic elements of sea urchins. (*Strongylocentrotus*). III. *In vitro* reaction to bacteria. *J. Invert. Pathol.* **13,** 42-62.

JOHNSON, P. T. and CHAPMAN, F. A., 1970. Infection with diatoms and other micro-organisms in sea urchin spines (*Strongylocentrotus francescanus*). *J. Invert. Pathol.* **16,** 268-276.

LASHLEY, B. J. and NIGRELLI, R. F., 1970. The effect of crude holothurin on leucocyte phagocytosis. *Toxicon* **8,** 301-306.

LEWIS, J. B. and SALUJA, G., 1967. The claviform pedicellariae and their stalk glands in the tropical sea urchin *Diadema antillarum* Phillipi. *Can. J. Zool.* **45,** 1211-1214.

MACKIE, A. M., 1970. Avoidance reactions of marine invertebrates to either steroid glycosides of starfish or synthetic surface-active agents. *J. exp. mar. Biol. Ecol.* **5,** 63-69.

MACKIE, A. M., LASKER, R. and GRANT, P. T., 1968. Avoidance reactions of a mollusc *Buccinum undatum* to saponin-like surface active substances in extracts of the starfish *Asterias rubens* and *Marthasterias glacialis*. *Comp. Biochem. Physiol.* **26,** 415-428.

MACKIE, A. M. and TURNER, A. B., 1970. Partial characterization of a biologically active steroid glycoside isolated from the starfish *M. glacialis*. *Biochem. J.* **117,** 543-550.

MARGOLIN, A. S., 1964. The running response of *Acmaea* to sea stars. *Ecology* **45,** 191-193.

MILLOTT, N., 1969. Injury and the axial organ of Echinoids. *Experientia* **25,** 756-757.

MONTGOMERY, D. H., 1967. Responses of two haliotid gastropods (Mollusca) *Haliotis assimilis* and *H. rufescens* to the forcipulate asteroids (Echinodermata) *Pycnopodia helianthoides* and *Pisaster ochraceous*. *Veliger* **9,** 359-368.

POPE, E., 1968. Venomous starfish in Sydney harbour. *Aust. nat. hist.* **16,** 26.

ROSS, D. M. and SUTTON, L., 1967. Swimming sea anemonies of Puget Sound: Swimming of *Actinostola*, a new species, in response to *Stomphia coccinea*. *Science, N.Y.* **155,** 1419-1421.

RUSSELL, F. E., 1967. Comparative pharmacology of some animal toxins. *Fed. Proc.* **26,** 1206-1224.

SHIMADA, S., 1969. Anti-fungal steroid glycoside from sea cucumber. *Science, N.Y.* **163,** 1462.

SMITH, G. N., 1971a. Regeneration in the sea cucumber *Leptosynapta*. I. The process of regeneration. *J. exp. Zool.* **177,** 319-330.

SMITH, G. N., 1971b. Regeneration in the sea cucumber *Leptosynapta*. II. The regenerative capacity. *J. exp. Zool.* **177,** 331-342.

SNYDER, N. and SNYDER, H., 1970. Alarm response of *Diadema antillarum*. *Science, N.Y.* **168,** 276-278.

SOBOTKA, H., 1965. Comparative biochemistry of marine animals. *Bioscience* **15,** 583-584.

SOBOTKA, H., FRIESS, S. L. and CHANLEY, J. D., 1962. Physiological effects of holothurin, a saponin of animal origin. *Proc. 5th int. Neurochem. Symp.*, pp. 471-478.

THOMAS, G. E. and GRUFFYDD, Ll. D., 1971. The types of escape reactions elicited in the scallop *Pecten maximus* by selected sea star species. *Mar. biol. Berlin* **10,** 87-93.

WARD, J. A., 1965. An investigation on the swimming reaction of the anemone *Stomphia coccinea*. I. Partial isolation of a reacting substance from the asteroid *Dermasterias imbricata*. *J. exp. Zool.* **158,** 357-364.

WARD, J. A., 1968. An investigation on the swimming reaction of the anemone *Stomphia coccinea*. II. Histological location of a reacting substance in the asteroid *Dermasterias imbricata*. *J. exp. Zool.* **158,** 365-372.

WOBBER, D. R., 1970. A report on the feeding of *Dendronotus iris* on the anthozoan *Cerianthus* sp. from Monterey Bay, Calif. *Veliger* **12,** 383-387.

YASUMOTO, T. and HASHIMOTO, Y., 1965. Properties and sugar components of Asterosaponin A, isolated from starfish. *Agric. Biol. Chem.* **29,** 804-808.

YASUMOTO, T. and HASHIMOTO, Y., 1967. Properties of Asterosaponin B, isolated from a starfish *Asterias amurensis*. *Agric. Biol. Chem.* **31,** 368-372.

YASUMOTO, T., TANAKA, M. and HASHIMOTO, Y., 1966. Distribution of saponin in echinoderms. *Bull. Jap. Soc. scient. Fish.* **32,** 673-676.

YASUMOTO, T., WATANABE, T. and HASHIMOTO, Y., 1964. Physiological activities of starfish saponin. *Bull. Jap. Soc. scient. Fish.* **30,** 357-364.

Chapter 7. Sensory Physiology

Part 1. Gravity, righting reaction and response to mechanical stimuli

PONAT, A., 1967. Untersuchungen zur zellulaeren Druckresistenz verschiedener Evertebraten der Nord und Ostsee. (Investigations on the cellular resistance to pressure in diverse bottom invertebrates of the North Sea and the Baltic Sea.) *Kiel. Meeresforsch.* **23,** 21-27.

SCHLIEPER, C., 1968. High pressure effects on marine invertebrates and fishes. *Mar. Biol. Berlin* **2,** 5-12.

Chapter 7. Sensory Physiology

Part 2. Chemical, thermal and photic stimuli

ANDERSON, H. A., MATHIESON, J. W. and THOMPSON, R. H., 1969. Distribution of spinochrome pigments in echinoids. *Comp. Biochem. Physiol.* **28,** 333-345.

BULLOCK, E. and DAWSON, C. J., 1970. Carotenoid pigments of the holothurian *Psolus fabrichii* Duben & Koren. *Comp. Biochem. Physiol.* **34,** 799-804.

CAMPBELL, A. C. and LAVERACK, M. S., 1968. The responses of pedicellariae from *Echinus esculentus* L. *J. exp. mar. Biol. Ecol.* **2,** 191-214.

CHIA, F.-S., 1968. Histology of the pedicellariae of the sand dollar *Dendraster excentricus. J. Zool. Lond.* **157,** 503-507.

CHIA, F.-S., 1969. Response of the globiferous pedicellariae to inorganic salts in three regular echinoids. *Ophelia* **6,** 203-210.

CHIA, F.-S., 1970. Histology of the globiferous pedicellariae of *Psammechinus miliaris. J. Zool. Lond.* **160,** 9-16.

DAMBACH, M., 1969. Die Reaktion der Chromatophoren des Seeigels *Centrostephanus longispinus* auf Licht. *Z. vergl. Physiol.* **64,** 400-406.

DAMBACH, M. and HENTSCHEL, G., 1970. The covering reactions in sea urchins: New experiments and interpretations. *Mar. Biol. Berlin* **6,** 135-141.

DAMBACH, M. and JOCHUM, F., 1968. Zum verlauf der Pigmentausbreitung beim Farbwechsel des Seeigels *Centrostephanus longispinus* Peters. *Z. vergl. Physiol.* **59,** 403-412.

DIX, T. G., 1970. Covering response of the echinoid *Evechinus chloroticus. Pac. Sci.* **24,** 187-194.

FRICKE, H.-W. VON and HENTSCHEL, M., 1971. Die Garnelen-Seeigel—Partnershaft; eine Untersuchung der optischen Orientierung der Garnele. *Z. Tierpsychol.* **28,** 453-462.

GOUGH, J. H. and SUTHERLAND, M. D., 1970. Pigments of marine animals. IX. A synthesis of 6-acetyl 2.3.5.7. tetrahydroxy 1 : 4 naphthoquinone, its status as an echinoid pigment. *Aust. J. Chem.* **23,** 1839-1846.

HOLLAND, N. D., 1967. Some observations on the saccules of *Antedon medit-terranea* (Echinodermata: Crinoidea). *Publ. Staz. Zool. Napoli* **35,** 257-262.

HORA, J., TOUBE, T. P. and WEEDON, B. C. L., 1970. Carotenoids and related compounds. XXVII. Conversion of fucoxanthin to paracentrone. *J. Chem. Soc.* (Sect. C) **2,** 241-242.

KENT, R. A., SMITH, I. R. and SUTHERLAND, M. D., 1970. Pigments of marine animals. X. Substituted naphthopyrones from the crinoid *Comantheria perplexa*. *Aust. J. Chem.* **23,** 2325-2335.

MARTIN, R. B., 1968. Aspects of the ecology and behaviour of *Axiognathus squamata* (Echinodermata: Ophiuroidea). *Tane* **14,** 65-81.

MATHIESON, J. W. and THOMSON, R. H., 1971. Naturally occurring quinones. XVIII. New spinochromes from *Diadema antillarum, Spatangus purpureus* and *Temnopleurus toreumaticus*. *J. Chem. Soc.* (Sect. C) **1,** 153-160.

OVIATT, C. A., 1969. Light influenced movement of the starfish *Asterias forbesi* (Desor). *Behaviour* **33,** 52-57.

RAO, K. R. and FINGERMAN, M., 1970. Chromatophorotropic activity of extracts of radial nerves from starfish *Asterias amurensis*. *Z. vergl. Physiol.* **67,** 133-139.

SALAQUE, A., BARBIER, M. and LEDERER, F., 1967. Sur la biosynthèse de l'échinochrome A par l'oursin *Arbacia pustulosa*. *Bull. Soc. Chim. biol.* **49,** 841-848.

SALAZAR, M. H., 1970. Phototaxis in the deep sea urchin *Allocentrotus fragilis* (Jackson). *J. exp. mar. Biol. Ecol.* **5,** 254-264.

SINGH, H., MOORE, R. E. and SCHEUER, P. J., 1967. The distribution of quinone pigments in echinoderms. *Experientia* **23,** 624-626.

YOSHIDA, M. and OHTSUKI, H., 1968. The phototactic behaviour of the starfish *Asterias amurensis* Lutken. *Biol. bull. mar. biol. Lab. Woods Hole* **134,** 516-532.

Chapter 8. Physiology of the Water Vascular System and the Neural Control of Locomotion

ANSELL, A. D., 1967. Leaping movements in two species of Asaphidae. *Proc. Malac. Soc.* **37,** 395-398.

BARNES, D. J., BRAUER, R. W. and JORDAN, M. R., 1970. Locomotor response of *Acanthaster planci* to various species of coral. *Nature, Lond.* **228,** 342-343.

HARRISON, G., 1968. Sub-cellular particles in echinoderm tube feet. *J. Ultrastruct. Res.* **23,** 124-133.

KAWAGUTI, S., 1965. Electron microscopy on the ampulla of the echinoid. *Biol. J. Okayama Univ.* **11,** 75-86.

LAFOND, E. C., 1967. Movements of benthonic organisms and bottom currents as measured from the bathyscaphe "Trieste". *Johns Hopkins Oceanogr. stud.* **3,** 295-301.

PARACER, S. M., 1967. Additional information concerning Tiedemann's bodies of *Asterias forbesi* (Desor). *Biol. bull. mar. biol. Lab. Woods Hole* **133,** 478.

DESOUSA SANTOS, H., 1965. Estudo da Ultrastrutura dos pés ambulacrários de *Asterina stellifera* (Hupé)—Echinodermata—Asteroidea. *Fac. Fil. Cienc. Letras Univ. Sao Paulo* **25,** 175-299.

DESOUSA SANTOS, H. and DESILVA SASSO, W., 1968. Morphological and histochemical studies on the secretory glands of starfish tube feet. *Acta Anat.* **69,** 41-51.

DESOUSA SANTOS, H. and DESILVA SASSO, W., 1970. Ultrastructural and histochemical studies on the epithelium revestment layer in the tube feet of the starfish *Asterina stellifera*. *J. Morphol.* **130,** 287-296.

Chapter 9. Physiology of Respiration

FARMANFARMAIAN, A., 1968. The controversial echinoid heart and hemal system—function effectiveness in respiratory exchange. *Comp. Biochem. Physiol.* **24,** 855-863.

HOLLAND, N. D., 1970. The fine structure of the axial organ of the feather star *Nemaster rubiginosa* (Echinodermata, Crinoidea). *Tissue Cell.* **2,** 625-636.

JOHANSEN, K. and PETERSEN, J. A. 1971. Gas exchange and active ventilation in a starfish *Pteraster tesselatus. Z. vergl. Physiol.* **71,** 365-381.

LEWIS, J. B., 1968. Comparative respiration of tropical echinoids. *Comp. Biochem. Physiol.* **24,** 649-652.

LEWIS, J. B., 1969. Respiration in the tropical sea urchin *Diadema antillarum* Phillipi. *Physiol. Zool.* **41,** 476-480.

MUELLER, W. E. G., ZAHN, R. K. and SCHMID, K., 1970. Morphologie und funktion der Cuvierschen Organe von *Holothuria forskali* Delle Chiaje. *Z. wiss. Zool.* **181,** 219-232.

OLAH, E. H. and ALLEMAND, B. H., 1966. Métabolisme oxydatif de quelques organismes récoltés en Meditterranée: Teneur en acid lactique, production anaerobic d'acid lactique, activité de la cytochromeoxydase et teneur en mucopolysaccharide. *Recl. Trav. Sta. mar. Endoume.* **57,** 3-8.

PENTREATH, R. J., 1971. Respiratory surfaces and respiration in three New Zealand intertidal ophiuroids. *J. Zool. Lond.* **163,** 397-412.

TERWILLIGER, R. C. and READ, K. R. H., 1970. The haemoglobins of the holothurian echinoderms *Cucumaria miniata* Brandt, *Cucumaria pipiata* Stimpson and *Molpadia intermedia* Ludwig. *Comp. Biochem. Physiol.* **36,** 339-351.

Chapter 10. The Physiology of Spawning and Neurosecretion

BRUSLE, J., 1969a. Les cycles génitaux d'*Asterina gibbosa* Pennant. *Cah. Biol. mar.* **10,** 271-287.

BRUSLE, J., 1969b. Radiosensibilité des cellules germinales et somatiques de la gonade hermaphrodite d'*Asterina gibbosa* Pennant après irradiation X: Étude ultrastructurale des radio lesions différentielles. *Arch. Biol.* **80,** 451-470.

CHIA, F.-S., 1966. Brooding behaviour of a six rayed starfish *Leptasterias hexactis. Biol. bull. mar. biol. Lab. Woods Hole* **130,** 304-315.

CHIA, F.-S., 1968. Some observations on the development and cyclic changes in the oocytes in a brooding starfish *Leptasterias hexactis. J. Zool. Lond.* **154,** 453-461.

DIX, T. G., 1969. Aggregation in the echinoid *Evechinus chloroticus. Pac. Sci.* **23,** 123-124.

FISHELSON, L., 1968. Gamete shedding behaviour of the feather star *Lamprometra klunzingeri* in its natural habitat. *Nature, Lond.* **219,** 1063.

HEILBRUNN, L. V., CHAET, A. B., DUNN, A. and WILSON, W. L., 1954. Antimitotic substance from ovaries. *Biol. bull. mar. biol. Lab. Woods Hole* **106,** 158-168.

HIRAI, S. and KANATANI, H., 1971. Site of production of meiosis inducing substance in the ovary of starfish. *Exptl. Cell Res.* **67,** 224-227.

HIRAI, S., KUBOTA, J. and KANATANI, H., 1971. Induction of cytoplasmic maturation by 1-methyl adenine in starfish oocytes after removal of the germinal vesicle. *Exptl. Cell. Res.* **68** 137-143.

IKEGAMI, S., SHIRAI, H. and KANATANI, H., 1971. Occurrence of progesterone in the ovaries of starfish. *Zool. Mag. Tokyo* **80,** 26-28.

KANATANI, H., 1969. Induction of spawning and oocyte maturation by 1-methyl adenine in starfishes. *Exptl. Cell. Res.* **57,** 333-337.

KANATANI, H., IKEGAMI, S., SHIRAI, H., OIDE, H. and TAMURA, S., 1971. Purification of gonad stimulating substance obtained from radial nerves of the starfish *Asterias amurensis. Dev. Growth Diff.* **13,** 151-164.

KANATANI, H., KUROKAWA, T. and NAKANISHI, K., 1969. Effects of various adenine derivatives on oocyte maturation and spawning in the starfish. *Biol. bull. mar. biol. Lab. Woods Hole* **137,** 384-385.

KANATANI, H. and SHIRAI, H., 1969. Mechanism of starfish spawning. II. Some aspects of action of a neural substance obtained from radial nerve. *Biol. bull. mar. biol. Lab. Woods Hole* **137,** 297-311.

KANATANI, H. and SHIRAI, H., 1970. Mechanism of starfish spawning. III. Properties and action of meiosis inducing substance produced in gonad under influence of gonad stimulating substance. *Dev. Growth Diff.* **12,** 119-140.

KANATANI, H. and SHIRAI, H., 1971. Chemical structural requirements for induction of oocyte maturation and spawning in starfishes. *Dev. Growth Diff.* **13,** 53-64.

KAWAGUTI, S., 1965. Electron microscopy of the ovarian wall of the echinoid with special references to its muscles and nerve plexus. *Biol. J. Okayama Univ.* **11,** 66-74.

KRISHNAN, S., 1968. Histochemical studies on reproductive and nutritional cycles of the holothurian *H. scabra. Mar. biol. Berlin* **2,** 54-65.

LUNDBLAD, G., 1967. A study of cathepsins in extracts of *Brissopsis* ovaries. *Abkiv. Kem.* **26,** 79-86.

MILEIKOVSKII, S. A., 1968. Razmnozhenie morskoi avezdy *Asterias rubens* L. v. Belom Barents evom, Norvezhskom i drugikh evropeiskikh moryakh. (Breeding of the starfish *Asterias rubens* L. in the White, Barents, Norwegian and other European Seas.) *Okeanol.* **8,** 693-704.

PEARSE, J. S., 1968. Patterns of reproductive periodicities in four species of Indo-Pacific echinoderms. *Proc. Indian Acad. Sci.* (Sect. B) **67,** 247-279.

SCHOENER, A., 1968. Evidence for reproductive periodicity in the deep sea. *Ecology* **49,** 81-87.

SCHUETZ, A. W., 1967. Variable sensitivity of starfish ovarian tissue to radial nerve factor. *Exptl. Cell. Res.* **48,** 183-186.

SCHUETZ, A. W., 1969. Chemical properties and physiological actions of a starfish radial nerve factor and ovarian factor. *Gen. Comp. Endocrinol.* **12,** 209-221.

SCHUETZ, A. W., 1971. Induction of oocyte maturation in starfish by 1-methyl adenine: Role of the ovarian wall. *Exptl. Cell. Res.* **66,** 5-10.

SCHUETZ, A. W. and BIGGERS, J. D., 1967. Regulation of germinal vesicle breakdown in starfish oocytes. *Exptl. Cell. Res.* **46,** 624-628.

SCHUETZ, A. W. and BIGGERS, J. D., 1968. Effects of calcium on the structure and functional response of the starfish ovary to radial nerve factor. *J. exp. Zool.* **168,** 1-10.

SHIRAI, H. and KANATANI, H., 1968. On the stability of neural substance responsible for gamete-shedding in starfish. *Zool. Mag. Tokyo* **77,** 128-130.

STRATHMANN, R. R. and SATO, H., 1969. Increased germinal vesicle breakdown in oocytes of the sea cucumber *Parastichopus californicus* induced by starfish radial nerve extract. *Exptl. Cell. Res.* **54,** 127-129.

TANGAPREGASSOM, A. M. and DELAVAULT, R., 1967. Analyse en microscopie photonique et électronique des strictures périphériques des gonades chez deux étoiles de mer *Asterina gibbosa* et *Echinaster sepositus. Cah. biol. Mar.* **8,** 153-159.

Chapter 11. Physiology of Nerves and Muscles

BARGMANN, W. and BEHRENS, BR., 1968. Über die Pylorusanhänge des Seesternes (*Asterias rubens* L.) insbesondere ihre Innervation. *Z. Zellforsch mikrosk. Anat.* **84,** 563-584.

BARGMANN, W., VON HARNACK, M. and JACOB, K., 1962. Über den Feinbau des Nervensystems des Seesternes (*Asterias rubens* L.). I. Ein beitrag zur vergleichenden morphologie der glia. *Z. Zellforsch. mikrosk. Anat.* **56,** 573-594.

COBB, J. L. S., 1968a. The fine structure of the pedicellariae of *Echinus esculentus.* I. The innervation of the muscles. *J. R. microsc. Soc.* **88,** 211-221.

COBB, J. L. S., 1968b. The fine structure of the pedicellariae of *Echinus esculentus.* II. The sensory system. *J. R. microsc. Soc.* **88,** 223-233.

COBB, J. L. S., 1969. The distribution of monoamines in the nervous system of echinoderms. *Comp. Biochem. Physiol.* **28,** 967-971.

COBB, J. L. S., 1970. The significance of the radial nerve cords in asteroids and echinoids. *Z. Zellforsch. mikrosk. Anat.* **108,** 457-474.

COTTRELL, G. A. and LAVERACK, M. S., 1968. Invertebrate pharmacology. *Ann. Rev. Pharmacol.* **8,** 273-298.

COTTRELL, G. A. and PENTREATH, V. W., 1970. Localization of catecholamines in the nervous system of starfish *Asterias rubens* L. and a brittle star *Ophiothrix fragilis. Comp. Gen. Pharmacol.* **1,** 73-81.

COTTRELL, G. A. and WELSH, M., 1966. Effects of synthetic eledoisin and bradykinin on certain invertebrate preparations and the isolated frog heart. *Nature, Lond.* **212,** 838-839.

FEDER, H. M. and LASKER, R., 1968. A radula muscle preparation from the gastropod *Kelletia kelletii* for biochemical assays. *Veliger* **10,** 283-285.

VON HEHN, G., 1970. Ueber den Feinbau des hyponeuralen Nervensystems des Seesternes (*Asterias rubens* L.). *Z. Zellforsch. mikrosk. Anat.* **105,** 137-154.

HILL, R. B., 1970. Effects of some postulated neurohumours on rhythmicity of the isolated cloaca of a holothurian. *Physiol. Zool.* **43,** 109-123.

KAWAGUTI, S., KAMISHIMA, Y. and KOBAYASHI, K., 1965. Electron microscopy of the radial nerve cord of the sea urchin. *Biol. J. Okayama Univ.* **11,** 87-95.

LAVALLARD, R., SCHLENZ, R. and BALAS, G., 1965. Structures et ultrastructures du muscle protracteur de l'appareil masticateur chez *Echinometra lacunta* L. *Fac. Fil. Cienc. Letras Univ. Sao Paulo* **25,** 133-173.

MACKIE, G. O., SPENCER, A. N. and STRATHMANN, R., 1969. Electrical activity associated with ciliary reversal in an echinoderm larva. *Nature, Lond.* **223,** 1384-1385.

MENDES, E. G., ABBUD, L. and ANCONA-LOPEZ, A. A., 1970. Pharmacological studies on the invertebrate non-striated muscles. I. The response to drugs. *Comp. Gen. Pharmacol.* **1,** 11-22.

MOGNONI, G. A. and LANZAVECCHIA, G., 1969. Studi sulla muscolatura eliocoidale e paramiosinica. III. Osservazioni comparative sulle protein muscolari di mitilo e oloturia. *Atti Acad. naz. Lincei Rend. Cl. Sci. Fish Mat. Natur. Ser. III* **46,** 610-618.

PENTREATH, V. W. and COTTRELL, G. A., 1968. Acetylcholine and cholinesterase in the radial nerve of *Asterias rubens* L. *Comp. Biochem. Physiol.* **27,** 775-785.

SAITA, A., 1969. La morphologia ultrastrutturale dei muscoli della "Lanterna di Aristotele" di alcuni echinoidi. *Lombardo Accad. Sci. Lett. Rend. Sci. Biol., Med. B* **103,** 297-313.

SOKOLOV, V. A., 1966. Temporary connections in starfish. *Ref. Zh. Biol.* 1966, No. 11, D 332.

SUZUKI, R., KATSUMO, I. and MATANO, K., 1971. Dynamics of "Neuron ring": Computer simulation of central nervous system of starfish. *Kybernetik* **8,** 39-45.

WELSCH, U., 1971. Catecholamin und 5-Hydroxytryptamin im Nervensystem von Seesternen und Enteropneusten. *Z. Zellforsch. mikrosk. Anat.* **115,** 88-93.

The following papers also contain information of physiological interest.

BAKUS, G. J., 1968. Defensive mechanisms and ecology of some tropical holothurians. *Mar. biol. Berlin* **2,** 23-32.

BARGMANN, W. and BEHRENS, BR., 1964. Uber die Tiedemannschen organe des Seesternes (*Asterias rubens* L.). *Z. Zellforsch. mikrosk. Anat.* **63,** 120-133.

BARGMANN, W. and VON HEHN, G., 1968. Uber das Axialorgan ("mysterious gland") von *Asterias rubens* L. *Z. Zellforsch. mikrosk. Anat.* **88,** 262-277.

BARNES, J. H., 1966. The crown of thorns starfish as a destroyer of coral. *Aust. nat. Hist.* **15,** 257-261.

BRUN, E., 1968. Extreme population density of the starfish *Asterias rubens* L. on a bed of Iceland scallops *Chlamys islandica* (O. F. Muller). *Astarte* **32,** 1-3.

CHIA, F.-S., 1969. Some observations on the locomotion and feeding of the sand dollar *D. excentricus* (Eschschaltz). *J. exp. mar. biol. Ecol.* **3,** 162-170.

CAMPBELL, A. C. and ORMOND, R. F. G., 1970. The threat of the crown o' thorns starfish (*Acanthaster planci*) to coral reefs in the Indo-Pacific area: Observations on a normal population in the Red Sea. *Biol. Conserv.* **2,** 246-252.

COBB, J. L. S., 1969. The innervation of the oesophagus of the sea urchin *Heliocidaris erythrogramma. Z. Zellforsch. mikrosk. Anat.* **98,** 323-332.

COLEMAN, R., 1969a. Ultrastructure of the tube foot sucker of a regular echinoid *Diadema antillarum* Phillipi with special reference to secretory cells. *Z. Zellforsch. mikrosk. Anat.* **96,** 151-161.

COLEMAN, R., 1969b. Ultrastructure of the tube foot wall of a regular echinoid *Diadema antillarum* Phillipi. *Z. Zellforsch. mikrosk. Anat.* **96,** 162-172.

CRUMP, R. G., 1971. Annual reproductive cycles in three geographically separated populations of *Patiriella regularis* (Verrill), a common New Zealand asteroid. *J. exp. mar. biol. Ecol.* **7,** 137-162.

DANA, T. F., 1970. Acanthaster: A rarity in the past? *Science, N.Y.* **169,** 894.

DANA, T. and WOLFSON, A., 1970. Eastern Pacific crown o' thorns starfish populations in the lower Gulf of California. *Trans. San Diego soc. natur. Hist.* **16,** 83-90.

Devaney, D. M., 1967. An ecto-commencal polynoid associated with Indo-Pacific echinoderms, primarily ophiuroids. *Occas. pap. Bernice P. Bishop Mus.* **23,** 287-304.

Dix, T. G., 1970. Biology of *Evechinus chloroticus* (Echinoidea: Echinometridae) from different localities. 2. Movement. *N.Z. Jl mar. freshwater Res.* **4,** 267-277.

Ebert, T. A., 1967a. Growth and repair of spines in the sea urchin *Strongylocentrotus purpuratus* (Stimpson). *Biol. bull. mar. biol. Lab. Woods Hole* **133,** 141-149.

Ebert, T. A., 1967b. Negative growth and longevity in the purple sea urchin *Strongylocentrotus purpuratus* (Stimpson). *Science, N.Y.* **157,** 557-558.

Ebert, T. A., 1968. Growth rates of the sea urchin *Strongylocentrotus purpuratus* related to food availability and spine abrasion. *Ecology* **49,** 1075-1091.

Ebling, F. J., Dawkins, A. D., Kitching, J. A., Muntz, L. and Pratt, V. M., 1966. The ecology of Lough Ine. XVI. Predation and diurnal migration in the *Paracentrotus* community. *J. anim. Ecol.* **35,** 559-566.

Feder, H. M., 1970. Growth and predation by the Ochre star *Pisaster ochraceous* (Brandt) in Monterey Bay, California. *Ophelia* **8,** 161-185.

Gray, I. E., McCloskey, L. R. and Weihe, S. C., 1968. The commencal crab *Dissodactylus mellitae* and its reaction to sand dollar host factor. *J. Elisha Mitchell scient. Soc.* **84,** 472-481.

Grodzinska, N., 1970. Inwazja na Pacyfiku. (Invasion in the Pacific.) *Wszechswiat.* **1,** 14-16.

Hinegardner, R. T., 1969. Growth and development of the laboratory cultured sea urchin. *Biol. bull. mar. biol. Lab. Woods Hole* **137,** 465-475.

Jensen, M., 1969. Age determination in echinoids. *Sarsia* **37,** 41-44.

Johnson, I. S., 1952. The demonstration of a "host factor" in commencal crabs. *Trans. Kansas Acad. Sci.* **55,** 458-464.

Kawaguti, S., 1966. Electron microscopy on the body wall of the sea cucumber with special attentions to its mucus cells. *Biol. J. Okayama Univ.* **12,** 35-45.

Lawrence, J. M. and Dawes, C. J., 1959. Algal growth over the epidermis of sea urchins spines. *J. Phycol.* **5,** 269.

Lawrence, J. M. and Ferber, I., 1971. Substrate particle size and the occurrence of *Lovenia elongata* (Echinodermata: Echinoidea) at Taba, Gulf of Eilat (Red Sea). *Israel J. Zool.* **20,** 131-138.

Leclerc, M. and Delavault, R., 1971. Présence de fibres nerveuses dans la paroi du sinus axial et dans la paroi coelomique chez *Asterina gibbosa* Pennant. *C. r. hebd. Seanc. Acad. Sci. Paris* **272,** 3311-3313.

Magnus, D. B. E., 1967. Ecological and ethological studies and experiments on the echinoderms of the Red Sea. *Stud. Trop. Oceanogr. Inst. Mar. Sci. Univ. Miami* **5,** 635-664.

Markel, K. and Maier, R., 1967. Beobachtungen an lochbewohnenden Seeigeln. *Natur. Mus.* **97,** 233-243.

Masse, H., 1970. Contribution a l'étude de la macrofaune de peuplements des Sables fins infralittoraux des Côtes de Provence. I. La Baie de Bandol. *Tethys* **2,** 783-820.

Menton, D. N. and Eisen, A. Z., 1970. The structure of the integument of the sea cucumber *Thyone briareus. J. Morphol.* **131,** 17-36.

Millott, N. and Coleman, R., 1969. The podial pit: A new structure in the echinoid *Diadema antillarum* Phillipi. *Z. Zellforsch. mikrosk. Anat.* **95,** 187-197.

NEWMAN, W. A., 1970. Acanthaster: A disaster. *Science, N.Y.* **167,** 1274-1275.

PETERSON, J. A., SAWAYA, P. and PIN-YI, L., 1965. Alguns aspectos de Ecologia de Echinodermata. *Ann. Acad. Brazil Cienc.* **37,** (Suppl.), 167-170.

PETTIBONE, M. H., 1969. Remarks on the North Pacific *Harmathoe tenebricosa* Moore, and its association with asteroids (Echinodermata: Asteroidea). *Proc. biol. Soc. Wash.* **82,** 31-42.

VINE, P. J., 1970. Densities of *Acanthaster planci* in the Pacific Ocean. *Nature, Lond.* **228,** 341-342.

WARNER, G. F., 1971. On the ecology of a dense bed of brittlestars *Ophiothrix fragilis. J. mar. biol. Ass. U.K.* **51,** 267-282.

WEBER, J. N. and WOODHEAD, P. M. J., 1970. Ecological studies of the coral predator *Acanthaster planci* in the South Pacific. *Mar. biol. Berlin* **6,** 12-17.

WICKLER, W. and SEIBT, U., 1970. Das verhalten von *Hymenocera picta* Dana, einer Seesterne fressenden Garnele (Decapoda: Natantia, Gnathophyllidae). (Behaviour of the starfish-killing shrimp *Hymenocera picta.*) *Z. Tierpsychol.* **27,** 352-368.

INDEX